Simone Cristina Caldato Silva

Cold pre-mixed flexible sidewalk with added tire rubber

Simone Cristina Caldato Silva

Cold pre-mixed flexible sidewalk with added tire rubber

Cold pre-mix with added rubber: an alternative for routine maintenance services in urban paving

ScienciaScripts

Imprint

Cover image: www.ingimage.com

This book is a translation from the original published under ISBN 978-3-330-77124-6.

Publisher:
Sciencia Scripts
is a trademark of
Dodo Books Indian Ocean Ltd. and OmniScriptum S.R.L publishing group

120 High Road, East Finchley, London, N2 9ED, United Kingdom
Str. Armeneasca 28/1, office 1, Chisinau MD-2012, Republic of Moldova, Europe
Managing Directors: Ieva Konstantinova, Victoria Ursu
info@omniscriptum.com

Printed at: see last page
ISBN: 978-620-8-63570-1

CONTENTS

INTRODUCTION

Materials are an integral part of human life, since products are made from materials that compete with each other to conquer current and future markets.

According to Bertollo (2000), the volume of tires discarded and stored in the cities of Araçatuba, Bauru, Botucatu, Campinas, Limeira, Piracicaba, Presidente Prudente, Ribeirâo Preto, Santos, Sâo José do Rio Preto, Sâo Paulo and Sorocaba, concluding that the average annual per capita generation of tires per inhabitant is approximately six million tires that are discarded annually in the state of Sâo Paulo. This figure may be underestimated, as it is lower than the averages in other countries. In the United States, for example, it is assumed that 188 million tires are disposed of inappropriately each year, which corresponds to 65 % of all tires (ANIP, 2007).

Aware of the worldwide problem of disposing of waste tires in the environment, the National Association of the Tire Industry (ANIP), which represents new tire manufacturers in the country, reached the milestone of 100 million tires disposed of in an environmentally correct manner in 2005. This corresponds to approximately 500,000 tons of old or unsalvageable tires that were collected and destroyed in an environmentally correct manner (SOARES, 2007).

In the 1980s, large tire fires in the United States caused air, water and soil contamination. With this problem, interest arose in the development of research aimed at reusing rubber from replaceable tires in engineering works. In 1991, the Intermodal Surface Transportation Efficiency Act obliged the State Departments of Transportation (DOT) and the Environmental Protection Agency (EPA), in cooperation, to develop further studies aimed at the use of tires in the construction of asphalt sidewalks (ODA, 2002). As a result, American states began to build asphalt sidewalks with additions of ground rubber from discarded tires: in 1994, around 5% of asphalt sidewalks were built with ground rubber from discarded tires; in 1995, 10%; in 1996, 15% and from 1997, 20% (ODA, 2002).

In Brazil, there is still no similar law requiring the use of a minimum percentage of discarded tires in paving works. However, progress has already been made with Resolution No. 258 of the National Environment Council (CONAMA) of August 26, 1999, which establishes the concept that the producer is responsible for the final destination of the product. For example, in the case of tires, the carcass would go back to the automaker.

According to Resolution No. 258, as of January 1, 2002, producers and importers of tires are obliged to collect and give environmentally appropriate final disposal to disposable tires in the national territory and as of 2005 the reduction of environmental liabilities began (CONAMA, 2002).

The reuse of unsalvageable tires is a very difficult challenge all over the world, given their peculiarities of degradability (around 600 years), quantity, volume and weight and, above all, the great difficulty of providing them with a new ecologically and economically viable destination (GRECA, 2003).

Recycling is an alternative to this environmentally friendly approach, which consists of reusing a specific waste product in a useful and economically viable way. In the case under study, rubber mixed with asphalt is not just an extra, inert product, placed just to fill the asphalt; in reality, rubber is recognized worldwide as a great asphalt enhancer.

According to Clerman (2004), in order to obtain rubber, the tire goes through a process called regeneration. In this process, the rubber is separated from the other components and the steel mesh and nylon are recovered as qualified scrap. After this separation, the tire is reduced to powder, because the vulcanization process has already taken place and it cannot be used as a substitute for raw rubber. Vulcanized rubber is a relatively inert product that is difficult to combine or bond with other substances.

There is a lot of scientific research into alternative uses for these tires, such as use in low-performance concrete, containment structures, production of rubber artifacts and drainage systems for landfills.

According to research by Clerman (2004), studies have been carried out on hot-mix asphalt concrete (HMA) with the addition of asphalt binder or polymer-modified asphalt with the addition of rubber. However, the use of hot-mix bituminous concrete has been replaced by asphalt emulsions, which make it possible to store production on site for future use and application.

According to the CNT (2007), 52% of freight in Brazil is transported by road. With the increase in the volume of transportation and traffic, as well as the economic resources available, it is interesting to look for alternatives that meet the design specifications and achieve a final product with satisfactory performance and that is also economically viable.

This has led to major economic and environmental concerns, as the disposal of tires has become a worldwide problem.

Thus, in search of an alternative to meet the Brazilian standards for asphalt mixture test results, this study aimed to verify a possible alternative for the application of cold premixed asphalt with the addition of tire rubber on low to medium traffic streets, evaluating the test results of the asphalt mix with the addition of ground rubber from used tires, with a reference mix without additions, comparing them with current specifications using the Marshall Test Method.

1 ASPHALT EMULSIONS

Asphalt emulsion is a colloidal dispersion of an asphalt phase in an aqueous phase (direct), or an aqueous phase dispersed in an asphalt phase (inverse), with the help of an emulsifying agent (DNER, 1996). They are usually produced mechanically using equipment with a high shear capacity, such as colloid mills.

The colloid mill, illustrated in figure 1, consists of a high-speed rotor that rotates between 1000 rpm and 6000 rpm on a stator. The spacing between the rotor and the stator is typically 0.25 mm to 0.50 mm, adjustable (PETROBRÂS and ABEDA, 2008).

Figure 1 - Colloid mill.

Source: PETROBRÂS and ABEDA, (2008).

In this equipment, the asphalt in its liquid (heated) state is dispersed in the aqueous medium by shearing into small droplets which are surrounded by emulsifying agents to form a stable emulsion, illustrated in figure 2 (Abeda, 2001). Thus, when two globules of asphalt try to approach each other, they are repelled.

Figure 2 - Colloid mill

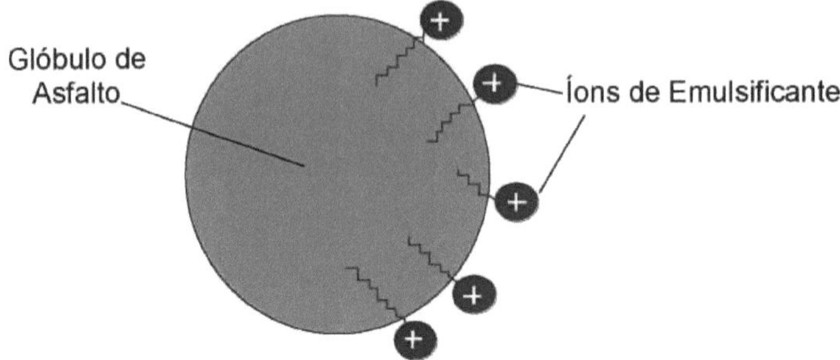

Source: ABEDA, (2001).

The sizes of the asphalt globules dispersed in water vary from 0.001 mm to 0.020 mm in diameter. The appearance varies from a low consistency liquid (e.g. milk) to a creamy consistency (e.g. honey), and the color of the emulsion is brown (ABEDA, 2001).

The asphalt globules are kept apart by the emulsifying agent until the emulsion is deposited either on the soil surface or on the existing sidewalk or aggregate particles (ASPHALT INSTITUTE, 1989).

The diameter of the particles in the asphalt emulsion used on roads is important because, with larger diameters, the emulsion is very coarse and tends to settle; with smaller diameters, stability is high and Brownian particle movement prevails.

According to Abeda (2001), industrial-scale production of asphalt emulsions began in Lutterbach in 1905 with chemist Emile Feigel. But it was in 1951 that Esso Brasileira de Petróleo in France took an important step in the use of this product, when it made cationic emulsions available to the market.

Currently, the world's largest consumer of asphalt emulsions for sidewalks is the United States with 2,300,000 metric tons per year, followed by France with 1,010,000 metric tons, corresponding to 27% of road services (Abeda, 2001). According to Santana (1993), Brazil's average annual production in recent years has been 750,000 tons of petroleum asphalt cement (PAC), around 30% of which is in the form of cationic emulsions (the only type specified and used in the country).

1.1 Emulsification process

In this process, petroleum asphalt cement (PAC) needs to be broken down into micrometric particles and dispersed in the aqueous medium.

In order for petroleum asphalt cement (PAC) to shear, i.e. to obtain the shear stress of PAC, thermal and mechanical energy must be applied through the colloid mill, where an emulsion of asphalt in water is obtained.

According to Abeda (2001), the asphalt cement is heated to a temperature of between 140 and 145 °C and the water phase is heated to a temperature of between 50 and 60 °C, in which the emulsifying agents are already dissolved, the main purpose of which is to prevent the asphalt particles from agglomerating, keeping the two phases in equilibrium for a period of time that can vary from a few weeks to a few months.

Emulsifying agents are substances added to emulsions to increase their kinetic stability, making them reasonably stable and homogeneous (ABEDA, 2001).

Figure 3 shows a simplified scheme for producing cationic asphalt emulsion.

Figure 3 - Basic scheme for producing cationic asphalt emulsion.

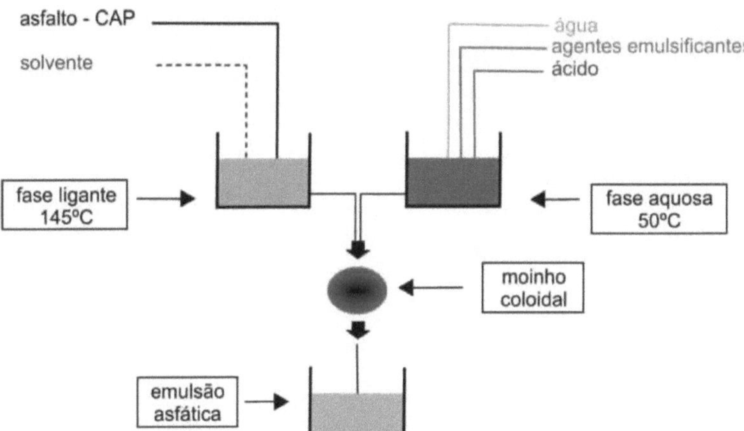

Source: ABEDA, (2001).

Emulsifying agents confer positive or negative electrical charges, and in some cases do not confer ionicity to emulsions and classify them as ationic, anionic and non-ionic (ABEDA, 2001).

1.2 Classification of Emulsions

Emulsions are classified according to the time required for the aqueous phase to separate from the asphalt phase (rupture), the asphalt content and the ionic charge.

Breaking occurs when the emulsion comes into contact with the aggregate. The speed at which this separation occurs depends (ABEDA, 2001):

- The type of emulsion;
- Reactivity in contact with the specific surface of the aggregates;
- Moisture content;
- The temperature of the materials; e,
- The environment.

Figures 4, 5 and 6 illustrate schematically the effect of the rupture of the cationic emulsion when it comes into contact with the aggregate.

Figure 4 - The emulsion comes into contact with the aggregate.

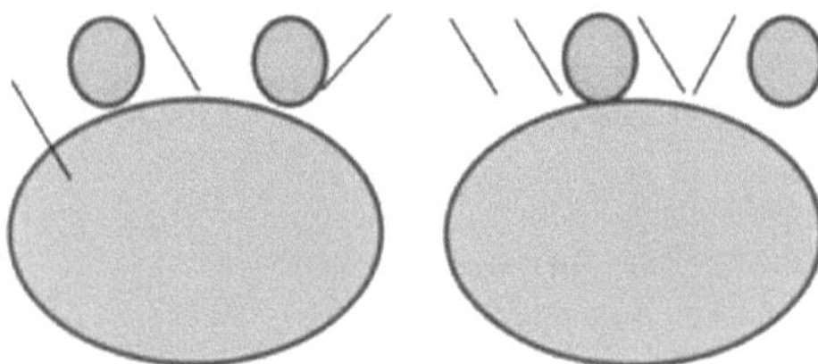

Source: ABEDA, (2001).

Figure 5 - Adsorption of the free emulsifying agent and attraction of the asphalt particles.

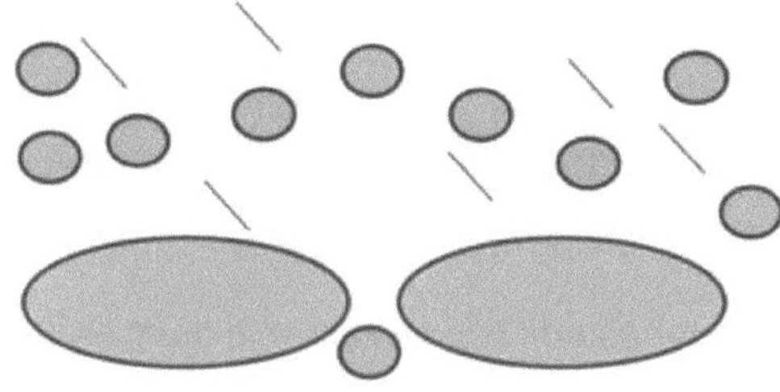

Source: ABEDA, (2001).

Figure 6 - Formation of the asphalt binder film on the aggregate.

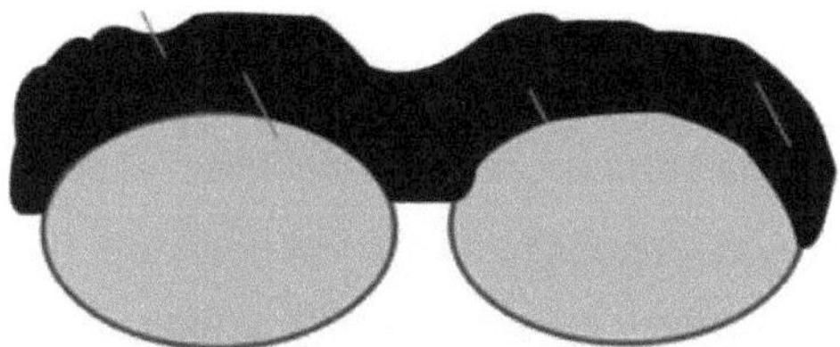

Source: ABEDA, (2001).

1.3 Types of emulsions

According to the DNER standard - EM 369 (1997) - Cationic Asphalt Emulsions, cationic asphalt emulsions have the symbols RR, RM and RL, followed by an indication and the letter C, according to their breakdown, "*Saybolt-Furol*" viscosity and solvent content.

The emulsion with the RR symbol is of the rapid type, breaking immediately or shortly after contact with the aggregates.

The emulsion with the RM symbol is of the medium type, its rupture occurs when the exposure time is longer than the rapid rupture. And the emulsion with the RL symbol is the slow type, its rupture occurs over a longer period than the two types, and it can be mixed with aggregates in the presence of fine material (stone dust).

Depending on the amount of asphalt cement involved in the manufacture of the emulsions, they can be classified as 1C and 2C, where the terminology C indicates the type of cationic emulsion and the numbers 1 and 2 are associated with the relative viscosity and the amount of asphalt cement used in the manufacture.

According to the Brazilian DNER standards, the classification of asphalt emulsions can be summarized into the following types and classifications:

a) RR-1C - a fast-breaking cationic asphalt emulsion with a minimum asphalt cement content of 62 % and a *Saybolt-Furol* viscosity (50 °C) of between 20 and 90 seconds, i.e. it has a low consistency. Its main field of application is in the service of painting connections between asphalt layers, followed by surface treatments by penetration;

b) RR-2C - a fast-breaking cationic asphalt emulsion with a minimum asphalt cement content of 67 % and a *Saybolt-Furol* viscosity (50 °C) of between 100 and 400 seconds, i.e. it has a high consistency. Its main field of application is in surface treatment services followed by the painting of bonds between asphalt layers;

c) RM-1C - medium-break cationic asphalt emulsion, with a minimum asphalt cement content of 62% and a "*Saybolt-Furol*" viscosity (50 °C) of between 20 and 200 seconds, i.e. it has a low consistency. Its main field of application is in the production of cold pre-mixed asphalt (PMF), followed by the painting of the bond between asphalt layers and cold sand-asphalt;

d) RM-2C - medium-break cationic asphalt emulsion, with a minimum asphalt cement content of 65 % and a viscosity [11]*Saybolt-Furol"* (50 °C) of between 100 and 400 seconds, i.e. it has a high consistency. Its main field of application is in the production of open pre-mixed asphalt (PMF), followed by the painting of asphalt-sand and cold asphalt-bonding layers;

e) RL-1C is a slow-setting cationic asphalt emulsion with a minimum asphalt cement content of 60% and a viscosity of [11]*Saybolt-Furol"* (50 °C) no

This means that it has a low consistency. Its main field of application is in the production of dense cold premix (PMF), followed by asphalt slurry, cold sand-asphalt and soil-emulsion stabilization;

f) LA-1C and LA-2C - cationic asphalt emulsion for asphalt slurry, which has a minimum asphalt cement content of 58 % and a viscosity of [11]Saybolt-Furol" (50 °C) at a maximum of 100 seconds, i.e. it has a low consistency. Its main field of application is in the production of asphalt slurry and soil-emulsion stabilization;

g) LA-E - non-ionic asphalt emulsion for asphalt slurry, with a minimum asphalt cement content of 58 % and a viscosity of *"Saybolt-Furol"* (50 °C) at a maximum of 100 seconds, i.e. it has a low consistency. Its main field of application is in the production of asphalt slurry and soil-emulsion stabilization;

h) Asphalt Emulsion with SBS Polymer (DNER-EM 396, 1999) - modified cationic emulsion containing at least 62 % polymer-modified asphalt (PMA) with SBS, which contains at least 3 % polymer by weight of PMA and SSF viscosity (25 °C) between 20 and 100 seconds. The polymer is incorporated directly into the asphalt at high temperatures before emulsification under specific technological conditions. Its main field of application is in the production of cold asphalt micro surfacing with SBS polymer-modified asphalt emulsion, according to DNER specifications;

i) Asphalt Emulsion with SBR Polymer (DNER-EM 396, 1999) - cationic modified emulsion, containing at least 62 % SBS polymer-modified asphalt (AMP) with at least 3 % polymer by weight of AMP and SSF viscosity (25 °C) between 20 and 100 seconds.

Its main field of application is in the production of cold asphalt micro *surfacing* according to the specifications of the ISSA - *International Sturry Surfacing Association*;

j) Asphalt emulsion with SBS polymer for bonding paint (DNER-ES 395, 1999) - Cationic emulsion modified by SBR or SBS polymer, containing at least 50% asphalt modified by polymer and SSF viscosity (25 °C) between

20 and 100 seconds. Used for bonding paint on the primed surface, or bituminous coating prior to the execution of any bituminous layer; e,

k) Emulsions for Cold Recycling - these are cationic emulsified recycling agents (with the exception of ARE-1, for which the particle load test does not apply), with at least 60% recycling agents, whether or not they contain polymers, and a SSF viscosity (25 °C) of no more than 70 seconds. It is used in the cold recycling of asphalt coatings to recover the original physico-chemical characteristics of the asphalt binder and formulated according to the project, specific to each job.

Asphalt emulsions must be homogeneous, with no separation of the asphalt phase after vigorous agitation within thirty days of loading.

Asphalt emulsions must meet the specific conditions shown in tables 1 and 2.

Table 1 - Cationic Emulsion Specifications - Tests on the emulsion.

Features	Test Methods	Types				
		Fast Breaking		Average Rupture		Slow rupture
		1C	2C	1C	2C	1C
a) Saybolt-Furol viscosity: SSF at 50° C	NBR 14491	20-90	100-	20-200	100-400	Max. 70
(b) Sedimentation, 5 days, % by weight maximum by difference	DNER ME 006	5	5	5	5	5
(c) Sieving (retained on 0.84 mm sieve) Maximum % by weight	DNER ME 005	0,1	0,1	0,1	0,1	0,1
d1) Water resistance, minimum % coverage: Dry aggregate	DNER ME 059	80	80	80	80	80
d2) Water resistance, minimum % coverage: Wet aggregate		80	80	60	60	60
e) Mixture with cement % maximum	DNER ME 007	-	-	-	-	2
or mix with silicon filer	DNER ME 008	-	-	-	-	1,2-2,0
f) Particle charge	DNER ME 002	positive	positive	positive	positive	positive
g) pH, maximum	DNER ME 149	-	-	-	-	6,5
(hi) Distillation: Distilled solvent, % by volume of total emulsion.	NBR 6568	0-3	0-3	0-20	0-12	-
(h2) Distillation: Distilled solvent, Residue, minimum % by weight		62	67	62	65	60
il) Demulsibility, minimum weight %	DNER ME 063	50	50	-	-	-

i2) Demulsibility, % maximum weight		-	-	50	50	-

Source: DNER EM 369, (1997).

Table 2 - Cationic Emulsion Specifications - Tests on waste.

Features	Test Methods	Types				
		Fast Breaking		Average Rupture		Slow rupture
		1C	2C	1C	2C	1C
a) Penetration at 25 °C, 100 g, 5s, 0.1 mm	DNER ME 003	50-250	50-250	50-250	50-250	50-250
b) Bitumen content, minimum % by weight	DNER ME 010	97	97	97	97	97
c) Ductility at 25°C, 5 cm/min, cm, minimum	DNER ME 163	40	40	40	40	40

Source: DNER EM 369, (1997).

Asphalt emulsions are used cold, providing logistical gains and reducing storage, application and transportation costs. Its use is compatible with practically all types of aggregates, obtaining excellent results. It can also be applied to wet aggregates, without the need for adhesive-improving additives (PETROBRAS, 2006).

According to Petrobràs (2006), checking the color of asphalt emulsions is a characteristic that becomes an auxiliary element for visual inspection and quick verification of the product's good condition. After breaking, the black color of the PAC prevails.

1.4 Advantages of cationic emulsions

According to Abeda (2001), the main advantages of cationic emulsions are presented below:

a) They represent an energy-saving alternative, since in most cases they can be used without the need for heating;

b) It has an excellent affinity with all types of aggregates, eliminating the use of "dope" additives, normally used to improve the adhesion of petroleum asphalt cement (PAC) in hot mixes with aggregates;

c) They make it possible to use wet aggregates, avoiding the need for fuel to dry them;

d) They can be stored at room temperature in simple installations that do not require a heating source, petroleum-based fuels or thermal insulation;

e) They eliminate the risk of fire and explosions, since no petroleum solvents are used;

f) They prevent the risk of burn accidents;

g) They do not generate toxic and polluting vapors, preserving the environment (environmentally friendly product); e,

h) They make it possible to produce large volumes of mixtures with low-cost equipment that is easy to operate and distribute.

In addition to these technical, operational, economic and environmental parameters, advances in asphalt modification technology have led to an increase in asphalt consumption worldwide, mainly in surface treatments, using emulsions with polymers to improve sidewalk performance in the face of traffic and climate.

2 COLD PREMIXED

Cold pre-mix is a mixture made at room temperature in an appropriate plant, made up of graded mineral aggregate, filler material (filer) and asphalt emulsion spread and compressed cold (DNER ES 317, 1997).

Cold ready-mix can be used as a coating, base, smoothing or reinforcement for sidewalks. It is easy to produce and is widely used by municipalities for paving streets in routine maintenance services, known as "pothole patching".

The constituent materials of cold premix are mineral aggregate and asphalt emulsion and must meet the specifications approved by the DNER - National Department of Highways.

The use of cold mixes has become increasingly widespread in technical circles around the world as an alternative solution for asphalt bases and overlays, thanks to their simplicity and speed (Brosseaud, 2002). However, in Brazil, its main use is still for the regularization layer and reinforcement of the sidewalk structure (binder) and routine maintenance services, such as "pothole patching" (ABEDA, 2001).

In the United States, to date, there has been little interest in using cold mixes with Asphalt Emulsion as an alternative to hot mixes in areas with heavy traffic or where hot mix plants are available. This is a consequence of the limitations of cold mixes with regard to climatic conditions and slow curing. In rural areas, where there are no hot-mix plants and where high transportation costs for fuel and other materials would be expected, asphalt-emulsion mixtures offer good advantages (AIKMAN, 2006).

The main factors to be evaluated that provide a suitable method for preparing and testing cold mixes with asphalt emulsion are: aggregate cover, workability in the mix and the moisture content retained in the samples before and after compaction (MAMLOUK *et al*, 1980).

In Brazil, according to Abeda (2001), the classification of Cold Pre-Mixed

Concrete (PMF) covers granulometry (amount of mortar and aggregates) and the amount of voids. Traditionally, PMF is classified into three classes according to its volume of voids (Vv), as can be seen in Table 3.

Table 3 - Classification of the PMF.

Name	% Empty
Open	22 a 30
Semi-dense	15 a 22
Dense	5 a 15

Source: ABEDA, (2001).

According to Santana (1993), the higher void volume values are due to two factors: cold mixes require more voids than hot mixes because, after compaction, the water and solvent (if any) must evaporate and, with the start of traffic, there will be a greater reduction in the void volume of cold mixes than hot mixes.

According to Abeda (2001), open-graded PMF can be stored for longer periods than dense-graded PMF (generally up to 30 days, as long as it is properly stored in piles and protected with waterproof tarpaulins). This characteristic provides flexibility in the scheduling of the work, meaning that priority can be given to the production phase of the asphalt mix and then to the application on the road. It also has the advantage of immediately releasing the asphalt layer to traffic, allowing the project to be built in stages.

Slow-breaking emulsions, on the other hand, are responsible for producing dense PMF, which looks like hot asphalt concrete (ABEDA, 2001).

Due to its low void content, this coating has excellent structural and functional performance for medium-volume roads, giving good performance both mechanically and in terms of surface finish, providing excellent rolling comfort (ABEDA, 2001).

According to DNER standards, among the various existing dosing methods,

Marshall dosing is the most widely used in Brazil (DNER ME 107, 1994). This method has been adapted for cold mixes and is known as Modified Marshall.

WALLER JR. (1980) recommends that the asphalt emulsion content used belongs to the following ranges:

- From 5 to 10 % for dense mixtures; e,
- From 5 to 8% for open mixtures.

The DNER ME 107 (1994) standard recommends wetting the aggregates before mixing them with the emulsion. This addition of water, called wetting moisture, helps to achieve good aggregate involvement (Santana, 1993) and prevents the discovery of fines (Mamlouk *et al*, 1980). In addition, the water used to wet the aggregate has a considerable effect on the workability of these mixtures (SILVA JUNIOR and SÓRIA, 2000).

Another recommendation from DNER 107 (1994) - Cold bituminous mix, with asphalt emulsion - Marshall test, is to cure between two and six hours before compaction for RR and RM emulsions and no more than one hour for RL emulsions.

The main disadvantage of using cold mixes with asphalt emulsions is the slow development of resistance, a phenomenon controlled by the loss of moisture in the mix (MAMLOUK *et al*, 1980).

The advantages of using cold mixes with asphalt emulsions are: easy storage, high bearing capacity and it is less aggressive to the environment as it does not require heating.

The use of cold mixes has become increasingly widespread in the field of materials technology. It is presented as an alternative solution in the execution of asphalt bases and coatings, mainly due to the speed of execution.

2.1 Aggregates

2.1.1 Bulk aggregate

According to DNER-ES 317 (1997), the coarse aggregate can be crushed stone, rolled pebbles, crushed or uncrushed, and must consist of sound, durable fragments, free of clay clods and harmful substances, and must have the following characteristics:

a) According to the DNER-ME 035 (1998) standard, Los Angeles wear must be equal to or less than 40 %, allowing for higher values if the aggregates used previously have performed satisfactorily;

b) According to DNER-ME 086 (1994), the shape index must be greater than 0.5%;

c) According to DNER-ME 089 (1994), the durability of aggregates must be less than 12 %;

d) According to the DNER-ME 083 standard (1994), the granulometry of the aggregates must comply with the ranges specified in Table 5,

e) According to the DNER-ME 059 (1994) standards, maintaining the adoption of the reference process of the ABNT NBR 6300 (2001) standard, adhesion must be greater than 90 %.

2.1.2 Small aggregate

The fine aggregate can be sand, stone dust or a mixture of both. Its individual particles must be resistant, have moderate angularity, be free of clods of clay and harmful substances and have a sand equivalent of 55% or more. (DNER-ME 054, 1997).

2.1.3 Filling material

It must be made up of finely divided, non-plastic materials, such as Portland cement, slaked lime, limestone powder, etc., and meet the following particle size requirements (table 4), when tested using the DNER-ME 083 (1994) method.

Table 4 - Granulometry of the filling material.

Sieves no.	mm	Minimum % passing
40	0,42	100
80	0,175	95
200	0,075	65

Source: DNER EM 083, (1994).

2.2 Composition of the mixture

According to DNER standard ES 317 (1997), the composition of the cold premix must meet the requirements, with the respective tolerances, with regard to grain size and bituminous binder percentages, as shown in Table 5.

Table 5 - Composition of the cold premix and its tolerances.

Sieves		Minimum % passing				
Pol.	mm	A	B	C	D	Design range tolerances
1"	25,4	100	-	100	-	±7%
3/4"	19,1	75-100	100	95-100	100	±7%
1/2"	12,7	-	75-100	-	95-100	±7%
3/8"	9,5	30-60	35-70	40-70	45-80	±7%
N°. 4	4,8	10-35	15-40	20-40	25-45	±5%
N°. 10	2,0	5-20	10-25	10-25	15-30	±5%
N°. 200	0,075	0-5	0-5	0-8	0-8	±2%
Bitumen soluble in CS2%		4-6	4-6	4-6	4-6	±2%

Source: DNER ES 317, (1997).

The Modified Marshall Method (DNER ME 107, 1994) should be used for cold mixes to check the conditions of voids, stability and creep, meeting the values shown in Table 6.

The Marshall Method is a method whose analysis ensures that an adequate volumetric proportion of the mixture's constituents is obtained, leading to a durable sidewalk.

The results obtained from this test are: Marshall stability and Marshall creep.

Marshall stability is defined as the maximum resistance to radial compression presented by the test specimen when molded and tested according to the process established in this method, expressed in Kgf and Marshall creep is defined as the total deformation presented by the test specimen, from the application of the initial zero load to the application of the maximum load, expressed in millimeters.

Table 6 - Conditions for cold mixing.

Percentage of Empties (%)	Stability Minimum (Kgf)		Marshall creep (mm)
5 a 30	250 (75 strokes)	150 (50 strokes)	2,0 - 4,5

Source: DNER ES 317, (1997).

2.3 Ground rubber

The process of vulcanizing rubber was discovered by chance by Charles Goodyear in 1839, when he dropped sulphur on a sample of rubber that was being heated. This led to the discovery of a more resistant and elastic material, the demand for which multiplied all over the world (GOODYEAR, 1999).

The idea of using tire rubber began in England around 1898, giving rise to a product called *"rubberbitumen"* (SPECHT, 2004). Since then, for years, engineers and chemists have tried to mix natural and synthetic rubber (polymers) with asphalt binders in order to improve their resilience properties. The growing interest in the use of rubbers has led to their use in paving, as well as in other alternatives, being studied and developed. However, many attempts to use this rubber have failed to produce exciting results.

Historically, the use of rubberized asphalt began in the 1940s, when the *U.S. Rubber Reclaiming Company* introduced to the market a product composed of asphalt material and recycled devulcanized rubber called "*Ramflex*" (GRECA, 2003).

Charles H. MacDonald is considered the father of rubberized asphalt in the United States. In 1963, he developed a highly elastic material to be used in the maintenance of asphalt sidewalks. The product was composed of asphalt binder and 25% ground tire rubber (0.6 to 1.2 mm), mixed at 190 °C for 20 minutes, to be used in patches, known as *"band-aid"* (GRECA, 2003). This *"band-aid"* was also used as a crack sealant (figure 7) and as a reinforcement layer (through the process called *Stress Absorbing Membrane Interlayer*, SAMI). The first application was made on a street in the city of Phoenix, where it could be seen that after six years, the sidewalk showed no reflection cracks. McDonald continued his experimental work in the city of Phoenix, together with the company *Atlos Rubber Inc.*, where sections were built at *Phoenix Sky Harbor International Airport* in 1966 (ODA, 2002).

Figure 7 - Crack in the sidewalk structure.

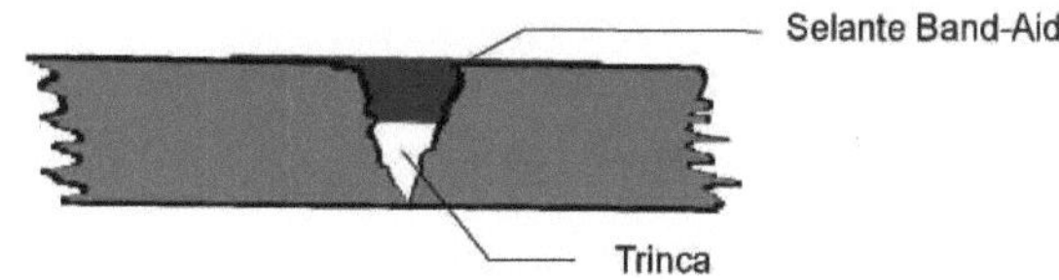

Source: Oda, (2002).

Since the 1980s, the addition of rubber from used tires to asphalt mixtures has also been considered an excellent alternative for reducing the environmental problems caused by solid waste disposal (figure 8). By the beginning of the 1990s, there were around 16,000 kilometers of roads built with asphalt-rubber. In 1991, the *Intermodal Surface Transportation Efficiency Act* (Section 1038 - *Public Law* 102-240) was formulated, which obliged the Departments of Transportation and Environmental Protection to develop studies to use rubber from discarded tires in paving materials (ODA, 2002).

Figure 8 - Millions of tires disposed of improperly.

Source: Oda, (2001).

According to data published by Echimenco (2001), at the beginning of this century there were around 100 million unserviceable tires abandoned in the country, either stored or dumped in open areas, to which he predicted around 17 million units would be added each year.

In 2006, according to ANIP - the National Association of the Tire Industry, Brazil produced 54.5 million tires and the sales volume reached 57.2 million tires, this figure includes the number of tires imported in 2006 (ANIP, 2007).

Currently, there is interest in asphalt mixtures, not with filler, but with rubber fragments, which consist of particles of vulcanized rubber. The vulcanization process converts a linear mass of molecules into a three-dimensional network, which gives objects their permanent shape (CLERMAN, 2004).

Tire rubber has excellent physical and chemical properties that can be incorporated into conventional binders, bringing a series of improvements that are directly reflected in the durability of the sidewalk: the incorporation of anti-oxidizing agents and inhibitors of the action of ultraviolet rays which significantly reduce the aging of the PAC, the increase in resistance to the chemical action of oils and fuels, the reduction in thermal susceptibility and the increase in admissible tensile deformation (improving fatigue behavior) (GRECA, 2003).

Since the 1960s, research laboratories and road agencies in different countries have been evaluating the application of these technologies. As a result, many ways of incorporating rubber into asphalt have been tested. According to Clerman (2004), conflicts between terminologies have caused difficulties in understanding and comparing data between road agencies.

According to GRECA (2003), tire rubber is most commonly added to asphalt mixtures in two ways:

- **Dry** rubber is fed directly into the asphalt plant's mixer. In this case, the rubber enters the mix as an aggregate. The transfer of important properties from the rubber to the binder is impaired, although it is possible to add improvements to the asphalt mix, as long as it is possible to obtain a homogeneous mix when manufacturing it; e,
- The rubber is previously mixed into the binder, permanently modifying it. In this way, the characteristics of elasticity and resistance to ageing are transferred more effectively to the original asphalt binder.

In the first case, the mixture is called Rubber-Aggregate or Rubber-Modified Asphalt Concrete, while in the second case it is called Rubber-Asphalt (HEITZMAN, 1992). Nowadays, a mixed process is also possible, where both categories are used. Table 7 summarizes these definitions.

Table 7 - Relationship between CRM terminologies.

Material	Process	Technology	Product
CRM	Wet	McDonald	Modified binder (rubber asphalt)
		Continuous Blending	
	Dry route	Plus Ride	Rubber Aggregate (rubber-modified asphalt concrete)
		Generic	
		Chunk Rubber	

Source: Adapted from Heitzman, (1992).

Table 8 shows the application of the rubber usage rate and its respective quantity.

Table 8 - Application of the rubber usage fee.

Place of Application	Quantity
Thin, crack-resistant coating with rubberized asphalt	0.2 kg rubber/ kg
Sealing cap with asphalt rubber	0.6 kg rubber/m^2
Binder-asphalt-rubber (wet)	5 % to 25 % on the percentage of asphalt binder
Rubber aggregate (dry)	3 % to 5 % ground rubber, in relation to the total mass of mineral aggregates

Source: Adapted from Oda et.al, (2005).

2.3.1 Wet process

In order to better understand the binder-asphalt-rubber interaction, we need some concepts that make it easier to understand the behavior of PAC based on its chemical composition. It is known that asphalt is a complex mixture of organic molecules that vary in chemical composition and molecular weight.

According to Clerman (2004), the chemical fractionation called SARA separates the components of PAC into four large groups:

- Saturated hydrocarbons: represent 5 to 20 % of asphalt and are the fraction least susceptible to oxidation;

- Aromatic hydrocarbons: have the lowest molecular weight and form the largest proportion of asphalt: 40 to 65 %. They have a low polarity and form a liquid that acts as a dispersant for asphaltenes;

- Resins: correspond to 5 to 30% of the PAC. Due to their high polarity, they are very adhesive, acting as a binding agent for the asphaltenes, thus increasing the fluidity of the asphalt; e,

- Asphaltenes: these are complex materials with a high molecular weight that typically make up 5 to 25 % of the PAC. Increasing the content of this fraction changes the consistency of the asphalt, producing a harder binder.

The first three fractions form the maltenes.

Rubber added to asphalt works like a polymer, absorbing liquids and swelling. The amount it absorbs depends on the type of polymer and the nature, temperature and viscosity of the asphalt, which acts as an organic solvent (TREOLAR, 1975).

As the rubber softens, the PAC will diffuse into each rubber particle, increasing the size of the rubber networks until equilibrium is reached. This elastomer, through its polymeric chains, absorbs all asphalt fractions with the exception of asphaltenes. This means that the maltenes are absorbed, leaving a high proportion of asphaltene in the residual asphalt, which has not diffused into the rubber, altering the rheological and cohesive properties of the residual asphalt (SINGLETON *et al*, 2000). Figure 9 shows the process.

Figure 9 - Schematic of the absorption of light asphalt fractions by rubber.

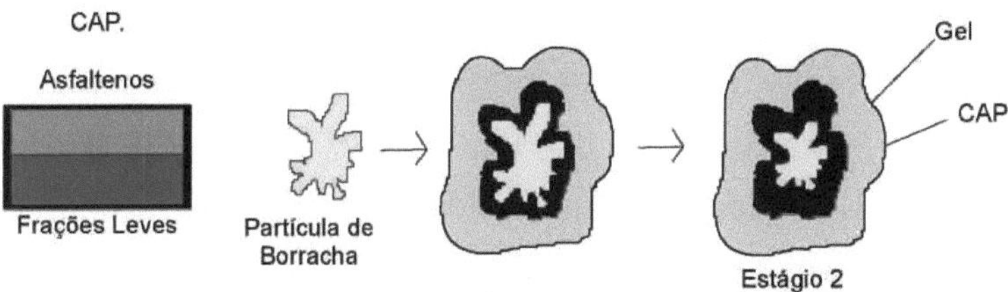

Stage 1

Source: Holleran and Reed, (2000) [2].

The reaction rate between this rubber and the asphalt can be accelerated by increasing the surface area of the rubber, which can be increased by reducing its particle size. Depending on the compatibility between these two materials, rubber, being an elastic material, makes it possible to modify the properties of the asphalt binder (HEITZMAN, 1992).

According to Way (2000), over the last 20 years, the Arizona Department of Transportation has found that asphalt-rubber mixtures, for their materials and climate, are excellent and durable surfacings. It has been observed:

- A significant reduction in crack reflections, even on roads with heavy and intense traffic;

- Noise reduction in urban areas; e,
- Excellent performance in terms of cracking and sinking of wheel tracks, leading to increased sidewalk durability.

In addition, rubberized asphalt has an excellent cost versus time ratio. These projects required less maintenance than those using conventional asphalt, even after ten years of service.

2.3.2 Dry process

For Heitzman (1992), the dry process is defined as any method of adding rubber directly to the hot mix, i.e. pre-mixing the rubber with the heated aggregate before mixing with the asphalt.

In the dry process, rubber is used as an aggregate and is a more recent alternative for the reuse of this material.

According to Airey *et al* (2003), the dry process has the potential to consume greater quantities of tire rubber than the wet process, resulting in greater environmental benefits. In addition, the production of dry mix asphalt is logistically simpler than wet mix.

According to Clerman (2004), the benefits of adding dry rubber, apart from environmental ones, are: increased flexibility, noise reduction, increased fatigue life and control of crack reflection.

Figure 10 illustrates the asphalt mix with rubber added as part of the aggregate.

Figure 10 - Illustration of rubberized asphalt modified by the dry process (the rubber is the particle in white).

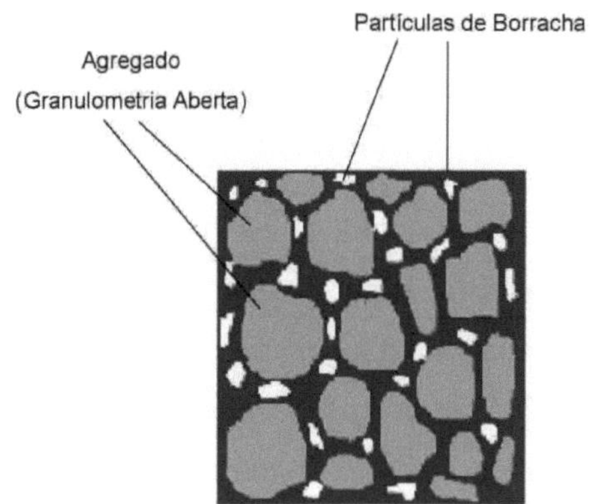

Source: Takallou and Hicks, (1988).

Experience in the state of Alaska has shown that an important factor in the success of hot dry mixes is a low void volume content. These vary from 0 to 4% depending on the level of traffic (NARUSCH, 1982).

In addition to the particle size and shape of the rubber particles, another aspect seems to be very important for dry mixes: the reaction time with the asphalt binder (CLERMAN, 2004).

The reaction time (or digestion time) of the rubber with the asphalt binder seems to be a very important parameter for the final performance of the mixture. Recent research highlights reaction times of between one and two hours as being adequate. During this period, the rubber can react with the asphalt, improving its final properties (PINHEIRO and SOARES, 2003).

In cold mixes, there is no consensus on the reaction time. Many authors believe that there is no possibility of a reaction between the rubber and the asphalt cement present in the asphalt emulsion. However, cold mixes already have a so-called curing time for compaction, where the rubber remains in contact with the PAC (CLERMAN, 2004).

SPECHT (2004) observed that the incorporation of dry rubber improves the characteristics of resistance to fatigue and thermal cracking, due to the lower sensitivity of the CBUQ to temperature variations. This author also carried out field tests, verifying through macrotexture readings that the rubber-aggregate surfacing has better functional characteristics than the comparative section in conventional asphalt concrete.

HUANG *et al* (2002) investigated the dry and wet processes by evaluating Marshall stability and creep, Modulus of Resilience (MR) and Tensile Strength (RT) tests and field performance for hot-mix asphalt mixes made in Louisiana, United States. According to the PMF classification, the rubberized open mixes (which have 22 to 30% voids) have higher creep values than the control mixes, while the dense mixes (which have 5 to 15% voids) have similar values. All the rubberized sections also showed greater deformations than the conventional coatings, indicating greater resistance to cracking. The analyses also showed much higher RT and MR values for the conventional mixes than for the rubberized mixes at temperatures of 5, 25 and 40 °C. The rubberized sections performed better in terms of wheel track subsidence (RTS), cracking, and irregularity index (IRI) than the corresponding sections even after five to seven years of traffic.

3 MATERIALS AND METHODS

A strip of cold premix was carried out, where the test specimens were produced in the laboratory with cationic asphalt emulsion. The addition of ground rubber and the levels of asphalt emulsion were defined and analyzed after the Marshall Cold Dosing test, according to DNER-ME 107 (1994).

The characteristics and behavior of the mixtures were evaluated by: asphalt mixture density, stability and Marshall creep.

All the tests were carried out at the Materials Testing Laboratory of the Technological Center of the Paulista Foundation, in the city of Lins/SP.

3.1 Materials

3.1.1 Asphalt emulsion

The asphalt emulsion used, supplied by a company in the state of São Paulo, was slow-breaking cationic asphalt emulsion (RL-1C).

Figure 11 shows the RL-1C emulsion storage tanks.

Figure 11 - Storage tanks.

Source: the author.

3.1.2 Aggregates

The aggregates used in this study were derived from basalt from a quarry near the city of Lins, São Paulo.

Figure 12 illustrates the storage of the aggregates: fine and coarse at the quarry. The aggregates were collected by a technician from the Materials Testing Laboratory of the Paulista Foundation Technology Center.

Figure 12 - Storage of large and small aggregates at the quarry.

Source: the author.

3.1.3 Rubber

Regenerating rubber involves separating it in its vulcanized state from the other components and digesting it with steam and chemicals, the product of which is refined in mills until a uniform blanket is obtained, or extruded to obtain granulated material. Regeneration is carried out by various degradation processes - alkaline, acid, mechanical and superheated steam. In regeneration, the waste undergoes modifications that make it more plastic and suitable for new vulcanization, but it does not have the same properties as the original raw rubber and is generally mixed with it to make artifacts. In the regeneration process used for tires, the rubber is separated from the other components, the wire and steel mesh are recovered as iron scrap and the nylon fabric is also recovered (NOHARA, *et al* 2005).

According to Greca (2003), a tire usually consists of:

- Structure made of steel, nylon, aramid fiber, rayon, fiberglass and/or polyester;

- Natural and synthetic rubber, incorporating hundreds of different types of polymer;
- Chemical reinforcers such as black carbon, silica and resins;
- Anti-degradants: paraffin waxes that are antioxidants and inhibit the action of ozone gas;
- Adhesion promoters: cobalt salts, metal baths on wires and resins;
- Curing agents: curing accelerators, activators, sulphur; e,
- Auxiliary products in tire processing such as oils.

A typical tire, such as the P195/75R14 model, for passenger cars has (GRECA, 2003):

- 2.50 kg of 30 different types of synthetic rubber;
- 2,05 kg of 8 different types of natural rubber;
- 2,27 kg of 8 types of black carbon;
- 0,68 kg of steel for the belt;
- 0,45 kg of polyester and nylon;
- 0,23 kg of steel wire; e,
- 1.36 kg of 40 different types of chemical components, waxes, oils, pigments, etc.

Of the approximate 9.54 kg of a tire, almost half is natural or synthetic rubber.

The ground rubber used in this work came from a company in the interior of the state of Sao Paulo. This rubber comes from discarded tires and is produced by swelling in a chemical organic product of aromatic origin followed by mechanical grinding, where the metals and fibers present in the tires are removed. The rubber is then fed into a high-pressure catalytic regeneration process. Shear rubber is obtained at the end of a drying process. The material obtained is a granular rubber (figure 13).

Figure 13 - Ground rubber.

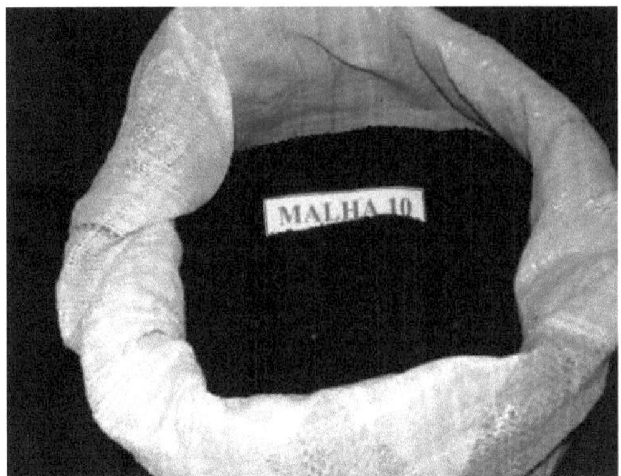

Source: the author.

3.2 Methods

The methodology used was based on experimentation to determine and analyze the various parameters involved in the process. A series of preliminary tests were carried out to compare the test results of an asphalt mixture with added tires with those of a reference mixture without added ground rubber.

The laboratory tests were carried out in accordance with the standards of the DNER - Departamento Nacional de Estradas e Rodagem published by the DNIT - Departamento Nacional de Infra- Estrutura de Transportes.

3.2.1 Tests

In compliance with the DNER ES-317 (1997) standard - Paving - cold pre-mix, here is a list of all the tests carried out, such as: granulometric analysis, Los Angeles abrasion, shape index, evaluation of durability, adhesiveness, Marshall test (stability and creep) and apparent density.

3.2.1.1 Aggregates

In sub-items: 4.2.1.1.1, 4.2.1.1.2, 4.2.1.1.3, 4.2.1.1.4 and 4.2.1.1.5, the tests carried out on the aggregates are described.

3.2.1.1.1 Particle size analysis

The aggregate granulometry tests were carried out in accordance with DNER standard ME 83 (1998) - Aggregates - Granulometric Analysis, which establishes the procedure for the analysis of fine and coarse aggregates by sieving.

The samples were first reduced in accordance with DNER PRO 199 (1996) and then dried in an oven at 110± 5 °C, cooled to room and their masses determined.

Sieving was carried out manually, initially on the sieve with the largest opening and then on the others in the series (in descending order).

The masses retained on each sieve, within the permitted tolerances, were applied in the calculations to obtain the results.

Figure 14 illustrates the particle size analysis test of the coarse aggregate (gravel).

Figure 14 - Example of a particle size analysis of coarse aggregate.

Source: the author.

3.2.1.1.2 Abrasion Los Angeles

The DNER ME 035 (1998) standard, which prescribes the method for determining the Los Angeles abrasion of aggregates, was used to carry out this test.

Los Angeles abrasion is the wear of the aggregate when it is placed in the Los Angeles machine together with an abrasive load and subjected to a certain number of revolutions of the machine at a speed of 30 rpm to 33 rpm (DNER ME 035, 1998).

After the particle size test and obtaining the grade of the sample in Band C, the mass of the dry sample was determined.

For this test, the abrasive load used was eight balls. The procedure for carrying out the test should be as follows: check the internal cleanliness of the drum, then place the sample and the respective abrasive charge, close the drum and rotate at a speed of 30 rpm to 33 rpm until 500 revolutions have been completed, remove all the material from the drum, separate the balls, clean and pass the sample through the 1.7 mm sieve, rejecting the material that has passed through. The material retained on the sieve was then washed and dried in an oven at between 105 and 110 °C for at least 3 hours. The material was then removed from the oven, cooled and the mass of the washed and dried sample was determined.

Figure 15 illustrates the equipment used to carry out the Los Angeles abrasion test.

Figure 15 - Equipment used for the Los Angeles abrasion test.

Source: the author.

1.1.1.1.3 shape index

The shape index test was carried out in accordance with DNER standard ME 086 (1994). This standard sets out the procedure for determining the shape index of aggregates and establishes the way in which the variation of the multidirectional axes of the particles that make up the aggregate is determined.

First, the material was analyzed in terms of particle size. Once the gradation had been chosen, the test was started using equipment called a reducing sieve.

The shape index result was obtained by adding the percentages retained in the sieves, divided by one hundred times the number of fractions that make up the chosen gradation.

Figure 16 shows the aggregate samples that were subjected to the shape index test, and figure 17 illustrates the necessary equipment used during the test.

Figure 16 - Example of aggregate for the shape index test.

Source: the author.

Figure 17 - Sieve used in the shape index test.

Source: the author.

1.1.1.1.4 Durability assessment

The standard used in this test was DNER ME 089 (1994) - Aggregates - durability assessment using sodium or magnesium sulphate solutions.

The method establishes how to determine the resistance to disintegration of aggregates subjected to the action of time, by the attack of saturated solutions of sodium or magnesium sulphate.

In the case under study, magnesium sulphate was used.

Before starting the test, a particle size analysis was carried out on a representative sample of the aggregate. The sample was then immersed in magnesium solution for 17 hours.

After the soaking period, the sample was removed from the solution, drained after 15 minutes and placed in an oven at 110 °C until it reached constant mass. During drying, precautions were taken to prevent the loss of particles retained in the sieves. After drying, the sample was cooled to room temperature.

The process of alternating soaking and drying constitutes a cycle, which must be repeated until the desired number of cycles has been completed.

After completing the final cycle and cooling the sample, the sample was washed in solution of solution so that it was free of excess magnesium sulphate. The sample was rinsed under running water. Freed from the solution, the sample fractions were dried until mass was constant. The material was then sieved as prescribed by the standard.

1.1.1.1.5 Adhesiveness

In 2001, the DNER ME 059 (1986) standard had to be adapted. Since then, the ABNT standard NBR 6300 (2001) - Cationic asphalt emulsions - determination of water resistance (adhesion) - has been adopted by reference.

Residual asphalt adhesion is the property of the aggregate being adhered by bituminous material, verified by the non-displacement of the asphalt film covering it, when the aggregate-residual asphalt mixture, after proper curing, is immersed in water at 40 °C for 72 hours.

For the emulsion under study (RL-1C), the sample can be tested at room temperature.

Here is a description of the procedure used for this test: first the sample is weighed and moistened with drinking water. It is then added to asphalt emulsion type RL-1C, until the entire surface is covered (25± 1) g. The sample is left to stand for 1 hour at room temperature, then taken to an oven at (60± 2) °C for 24 hours to cure completely. The sample is then transferred to the basket and allowed to come to room temperature. Then pour 400 ml of drinking water into the beaker and transfer the cooled sample to the beaker. Finally, the beaker with the sample is placed in an oven (40 ± 1) °C and kept at this temperature for 72 hours. The aggregate is then visually analyzed and the percentage of the area that remains covered by asphalt film is estimated.

3.2.1.2 Rubber

3.2.1.2.1 Particle size analysis

The rubber granulometry tests were also carried out in accordance with DNER standard ME 83 (1998) - Aggregates - Granulometric Analysis, which establishes the procedure for the analysis of fine and coarse aggregates by sieving.

Figure 18 illustrates an example of ground rubber with a mesh size of 10, which should be used for this study.

Figure 18 - Sample for the rubber particle size analysis test.

Source: the author.

3.2.1.3 Cold asphalt mix

The granulometric range for the asphalt mix chosen in this study is determined from the DNER ES 317 (1997) standard - Paving - Cold premixes. The working range adopted is Range C.

Tire rubber additions to asphalt mixtures vary from 3 to 5 % ground rubber in relation to the total mass of mineral aggregates (dry). This choice is based on research by ODA (2005).

Four traces were carried out for this study:

- Reference line - no ground rubber added;
- Trace 1 - addition of 3% ground rubber;
- Trace 2 - addition of 4% ground rubber; e,
- Trait 3 - addition of 5 % ground rubber.

3.2.1.3.1 Marshall essay

The DNER ME 107 (1994) standard - Cold bituminous mix with asphalt emulsion - Marshall test establishes how to determine the stability and creep of cold machined bituminous mix with asphalt emulsion, using the Marshall equipment.

According to the standard, five specimens should be molded per mix. To make the specimens, the percentages by weight in which the materials will be mixed must first be known. The quantity of each material is then calculated for a specimen weighing around 1200 g, with a height of 63.5 mm + 1.3 mm.

After the binder has been added, mixing must be rapid until the aggregates are completely covered. The material is then left to cure at room temperature for 60 minutes. After this time, compaction will begin.

After compaction, the specimens with the mold should be placed in an oven at 60 °C for 24 hours. After this period, they are removed from the oven until they reach room temperature.

The specimens are then removed from the molds and placed on a smooth, flat surface, taking care when handling them to avoid fracture or deformation. The specimens are then measured and weighed to determine the apparent density.

3.2.1.3.2 Apparent density

The DNER-ME 117 (1994) standard for Bituminous mixture - Determination of apparent density, establishes how to determine the density of bituminous mixtures in test specimens molded in the laboratory or obtained on the road (in loco).

Apparent density is the ratio between the weight of the mixture in air and the difference between the weight in air and the weight of the mixture suspended in water.

To carry out the test you will need: paraffin fluidized by heating, adhesive tape and a brush.

To begin the test, the specimens must be weighed, wrapped in adhesive tape and fluidized paraffin applied.

3.2.1.3.3 Marshall stability and fluency

The DNER ME 107 (1994) standard - Cold bituminous mix with asphalt emulsion - Marshall test establishes how to determine the stability and creep of cold machined bituminous mix with asphalt emulsion, using the Marshall equipment.

After the bulk density test, the specimens were dewaxed and placed in an oven at 40 °C for two hours.

They were then placed in the compression mold and taken to the press for breaking.

4 RESULTS AND DISCUSSIONS

4.1 Aggregates

Figure 19 illustrates the equipment used in the aggregate particle size test.

Figure 19 - Equipment used in the aggregate particle size analysis test.

Source: the author.

4.1.1 Particle size analysis

The materials were tested in accordance with DNER standard ME 83 (1998) - Aggregates - Particle size analysis.

Tables 9 and 10 show the results of the particle size analysis of the fine aggregates (pebbles and stone dust) and aggregates (12 and 16 mm gravel). The values shown are in percentages passing through the sieves.

Table 9 - Results of the granulometric analysis of the small aggregates.

Material		Gravel			Stone dust		
Sieves		Weight (g)		%	Weight (g)		%
mm	Pol.	Withheld	Passing	Passing	Withheld	Passing	Passing
50,0	2	0,0	2089,3	100,0	0,0	1059,4	100,0
38,0	1 1/2	0,0	2089,3	100,0	0,0	1059,4	100,0
25,4	1	0,0	2089,3	100,0	0,0	1059,4	100,0
19,1	3/4	0,0	2089,3	100,0	0,0	1059,4	100,0

12,5	1/2	0,0	2089,3	100,0	0,0	1059,4	100,0
9,5	3/8	0,0	2089,3	100,0	0,0	1059,4	100,0
4,8	N°. 4	817,3	1272,0	60,9	1,2	1058,2	99,9
2,0	N°.10	1201,4	70,6	3,4	124,9	933,3	88,1
0,42	N°. 40	9,7	60,9	2,9	492,0	441,3	41,7
0,18	N°. 80	7,1	53,8	2,6	126,7	314,6	29,7
0,075	N°. 200	12,9	40,9	2,0	118,4	196,2	18,5
Total (g)		2089,3	Total (g)		1059,4		

Source: the author.

With regard to the pebble and stone dust materials, 1272.0 g (60.9 %) and 1058.2 g (99.9 %) of material passed through the 4.8 mm sieve respectively.

Table 10 - Results of the particle size analysis of the coarse aggregates.

Material		16 mm gravel			16 mm gravel		
Sieves		Weight (g)		%	Weight (g)		%
mm	Pol.	Withheld	Passing	Passing	Withheld	Passing	Passing
50,0	2	0,0	5337,0	100,0	0,0	5164,0	100,0
38,0	1 1/2	0,0	5337,0	100,0	0,0	5164,0	100,0
25,4	1	0,0	5337,0	100,0	0,0	5164,0	100,0
19,1	3/4	0,0	5337,0	100,0	0,0	5164,0	100,0
12,5	1/2	1778,0	3559,0	66,7	0,0	5164,0	100,0
9,5	3/8	2919,8	639,2	12,0	364,8	4799,2	92,9
4,8	N°. 4	554,9	84,3	1,6	4674,3	124,9	2,4
2,0	N°.10	28,1	56,2	1,1	70,4	54,5	1,1
0,42	N°. 40	8,2	48,0	0,9	8,3	46,2	0,9
0,18	N°. 80	5,2	42,8	0,8	2,3	43,9	0,9
0,075	N°. 200	15,4	27,4	0,5	16,7	27,2	0,5
Total (g)		5337,0	Total (g)		5164,0		

Source: the author.

With regard to the 16 mm gravel material, the 12.5 mm sieve showed 3559.0 g (66.7%) of material passing through, while the 9.5 mm sieve showed 4799.2 g (92.9%) of material passing through.

The results shown in tables 9 and 10 comply with DNER standard 83 (1998).

4.1.2 Abrasion Los Angeles

Table 11 shows the results of the Los Angeles wear.

Table 11 - Results of the Los Angeles abrasion test on the aggregate.

Essay	Abrasion (%)	Average (%)
01	15,6	15,5
02	15,4	

Source: the author.

According to DNER standard ME 035 (1998), Los Angeles wear must be equal to or less than 40 %.

As shown in Table 11, the result of the Los Angeles abrasion test on the aggregate was 15.5% and complies with the DNER ME 035 (1998) standard.

4.1.3 shape index

The results of the shape index test are shown in Table 12.

Table 12 - Results of the aggregate shape index test.

<table>
<tr><th>Graduation</th><th colspan="2">Circular sieve (mm)</th><th>Weight</th><th colspan="4">Corresponding reduction sieves</th></tr>
<tr><th rowspan="3">D</th><th>Passing</th><th>Withheld</th><th>Initial Fraction</th><th>Sieve I (mm)</th><th>% Withheld</th><th>Sieve II (mm)</th><th>% Withheld</th></tr>
<tr><td>12,7</td><td>9,5</td><td>1000,5</td><td>6,3</td><td>57,0</td><td>4,2</td><td>30,2</td></tr>
<tr><td>9,5</td><td>6,3</td><td>1000,0</td><td>4,8</td><td>33,4</td><td>3,2</td><td>44,6</td></tr>
<tr><td colspan="4">Shape index</td><td colspan="4">0,64 %</td></tr>
</table>

According to DNER standard ME 086 (1994), the shape index must be greater than 0.5%.

Table 12 shows the results of the shape index, with a result of 0.64 %, in line with the DNER ME 086 (1994) standard, where the result was higher than 0.5 %.

4.1.4 Durability assessment

Figure 20 illustrates the materials that were tested in accordance with DNER standard ME 089 (1994) - Aggregates - Durability assessment using sodium or magnesium sulphate solutions.

Figure 20 - Aggregate samples for the durability assessment test.

For the durability test, DNER standard ME 089 (1994) recommends the use of sodium sulphate or magnesium sulphate.

In the case under study, magnesium sulphate was used.

Table 13 shows the results of the aggregate durability assessment.

Table 13 - Aggregate durability test results.

Particle size composition of the original sample			Test sample	% loss of each sample fraction		Weighted average
Passing mm	Withheld mm	% Withheld	Pasta Initial	Pasta Final	% Loss	Corrected Loss %
19,1	12,5	33,8	670,7	631,9	6,1	1,82
12,5	9,5	55,6	330,9	299,1	10,6	5,91
9,5	4,8	10,6	300,2	264,6	13,5	1,43

	Total	100,0	1301,8	1195,6	-	9,15
Total Loss				9,2 %		

Source: the author.

According to DNER standard ME 089 (1994), the durability of aggregates must be less than 12 %.

Table 13 shows that the total loss of aggregate was 9.2% and complies with DNER standard ME 089 (1994), which recommends a loss of less than 12%.

4.1.5 Adhesiveness

Figure 21 shows the aggregate samples after the adhesion test.

Figure 21 - Aggregate samples for the adhesion test.

Source: the author.

According to DNER standard ME 059 (1994), adhesion

must be higher than 90%.

The aggregate analyzed covered with asphalt film, after 72 hours of immersion, showed a visual estimate of 98 % adhesion to the emulsion used. This result complies with DNER standard ME 059 (1994).

4.2 Rubber

4.2.1 Particle size analysis

The materials were tested in accordance with DNER standard ME 83 (1998) -

Aggregates - Particle size analysis.

For this test, the sieving time was approximately two hours. This time depends on the type of material (elastic).

Table 14 shows the results of the particle size analysis of ground rubber with a mesh size of 10 (maximum characteristic size of 2 mm).

Table 14 - Particle size analysis of ground rubber.

Material		Rubber #10			Rubber #10		
Sieves		Weight (g)		%	Weight (g)		%
mm	Pol.	Withheld	Passing	Passing	Withheld	Passing	Passing
4,8	N°. 4	0,0	206,8	100,0	0,0	226,4	100,0
2,0	N°. 10	1,61	205,2	99,2	1,92	224,5	99,2
0,42	N° 40	203,66	1,6	0,8	222,46	2,0	0,9
0,18	N° 80	0,54	1,0	0,5	0,76	1,2	0,5
0,075	N° 200	0,54	0,5	0,2	0,72	0,5	0,2
4,8	N° 4	0,0	206,8	100,0	0,0	226,4	100,0
Total (g)		206,84	Total (g)		226,37		

Source: the author.

As for the ground rubber, 205.2 g (99.2 %) and 224.5 g (99.2 %) of the material passed through the 2.0 mm sieve.

The results shown in Table 14 comply with DNER standard ME 83 (1998).

4.3 Particle size composition of materials

Following the results of the material characterizations, the traits to be studied were defined.

These features are presented in sub-items: 4.3.1, 4.3.2, 4.3.3 e 4.3.4.

4.3.1 Reference line

Table 15 and figure 22 show the granulometric composition of the materials used in this work for the reference mix (without the addition of ground rubber).

This granulometric composition complies with the DNER-ES 317 (1997) standard - Paving - cold pre-mix.

The criterion for the adoption of this trait was based on the results obtained for the aggregates: fine and coarse, and ground tire rubber, i.e. the classification of the materials in grade C of DNER-ES 317 (1997).

Table 15 - Particle size composition of the reference mix materials.

Sieves		Material								Grade/Band: C		
		Brita 16mm		Brita 12mm		Gravel		Stone dust		Σ %	Specified Limits	
mm	Pol.	100%	35%	100%	30%	100%	15%	100%	20%	100%	Min.	Max.
25,4	1	100,0	35,0	100,0	30,0	100,0	15,0	100,0	20,0	100,0	100	100
19,1	3/4	100,0	35,0	100,0	30,0	100,0	15,0	100,0	20,0	100,0	95	100
9,5	3/8	12,0	4,2	92,9	27,9	10,0	1,5	100,0	20,0	53,6	40	70
4,8	N°. 4	1,6	0,6	2,4	0,7	60,9	9,1	99,9	20,0	30,4	20	40
2,0	N°.10	1,1	0,4	1,1	0,3	3,4	0,5	88,1	17,6	18,8	10	25
0,075	N°.200	0,5	0,2	0,5	0,2	2,0	0,3	18,5	3,7	4,3	0	8

Source: the author.

According to the results shown in Table 15, the particle size composition of the reference mix is as follows:

- 35 % 16 mm gravel;
- 30 % 12 mm gravel;
- 15% pebble; e,
- 20 % stone dust.

The sum of the percentages of these materials falls within band C (gradation) of the DNER-ES 317 (1997) standard - Paving - cold pre-mix.

Figure 22 shows an illustrative graph of the C range (working range), limited by the black lines, and the blue line shows the percentage sum of the reference line used in this study.

Figure 22 - Representation of the granulometric composition of the materials

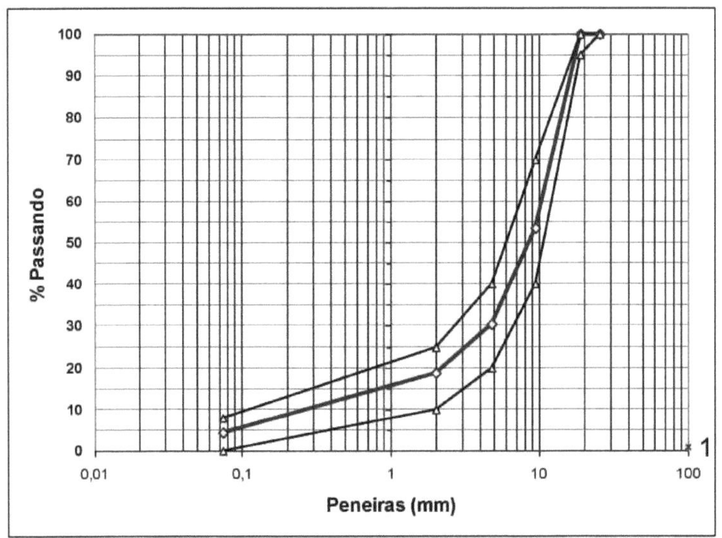

Source: the author.

4.3.2 Mix with 3 % ground rubber added

Table 16 and Figure 23 show the particle size composition of the materials used in the mix with the addition of 3% ground rubber.

According to the results shown in Table 16, the composite materials for this trait are:

- 35 % 16 mm gravel;
- 30 % 12 mm gravel;
- 15% pebbles;
- 20 % stone dust; e,
- 3 % ground rubber.

The sum of the percentages of these materials falls within band C (gradation) of the DNER-ES 317 (1997) standard - Paving - cold pre-mix.

Figure 24 shows a graph of range C (working range), limited by the black lines and the blue line shows the sum of the percentages of the mix with the addition of 3% ground rubber used in this study.

Table 16 - Particle size composition of the materials with the addition of 3% ground rubber.

Sieves		Material										Grade/Band: C		
		16mm gravel		12mm gravel		Gravel		Stone dust		Rubber		Σ %	Specific Limits cac	the
mm	Pol.	100%	35%	100%	30%	100%	15%	100%	20%	100%	3%	100%	Min.	Max.
25,4	1	100,0	35,0	100,0	30,0	100,0	15,0	100,0	20,0	100,0	3,0	103,0	100	100
19,1	3/4	100,0	35,0	100,0	30,0	100,0	15,0	100,0	20,0	100,0	3,0	103,0	95	100
9,5	3/8	12,0	4,2	92,9	27,9	10,0	1,5	100,0	20,0	100,0	3,0	56,6	40	70
4,8	N°. 4	1,6	0,6	2,4	0,7	60,9	9,1	99,9	20,0	100,0	3,0	33,4	20	40
2,0	N°.10	1,1	0,4	1,1	0,3	3,4	0,5	88,1	17,6	99,2	3,0	21,8	10	25
0,075	N°.200	0,5	0,2	0,5	0,2	2,0	0,3	18,5	3,7	0,2	0,0	4,3	0	8

Source: the author.

Figure 23 - Particle size composition of the materials with the addition of 3% ground rubber.

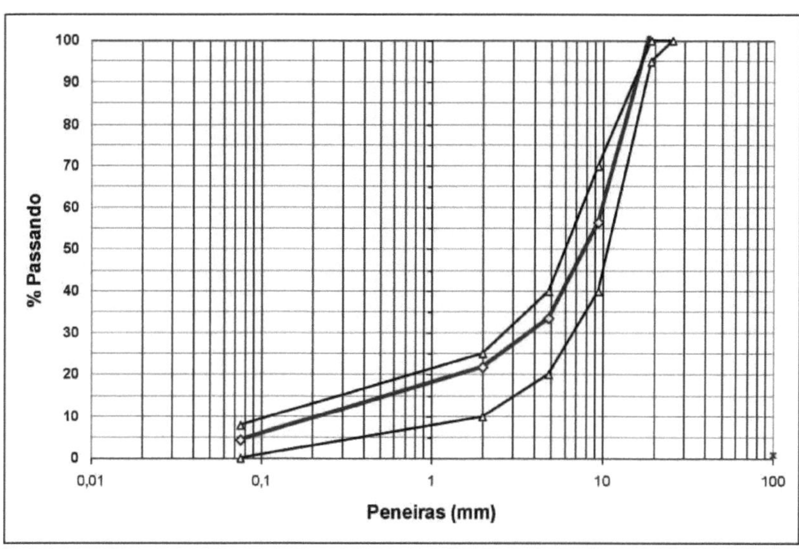

Source: the author.

4.3.3 Dash with 4 % ground rubber added

Table 17 and Figure 24 show the particle size composition of the materials used in the mix with the addition of 4% ground rubber.

According to the results shown in Table 17, the composite materials for this trait are:

- 35 % 16 mm gravel;
- 30 % 12 mm gravel;
- 20 % stone dust; e,
- 4 % ground rubber.

The sum of the percentages of these materials falls within band C (gradation) of the DNER-ES 317 (1997) standard - Paving - cold pre-mix.

Figure 25 shows a graph of range C (working range), limited by the black lines and the blue line shows the sum of the percentages of the mix with the addition of 4% ground rubber used in this study.

Table 17 - Particle size composition of the materials with the addition of 4% ground rubber .

Sieves		Material										Grade/Band: C		
		16mm gravel		12mm gravel		Gravel		Stone dust		Rubber		Σ %	Specific Limits cac	the
mm	Pol.	100%	35%	100%	30%	100%	15%	100%	20%	100%	4%	100%	Min.	Max.
25,4	1	100,0	35,0	100,0	30,0	100,0	15,0	100,0	20,0	100,0	4,0	104,0	100	100
19,1	3/4	100,0	35,0	100,0	30,0	100,0	15,0	100,0	20,0	100,0	4,0	104,0	95	100
9,5	3/8	12,0	4,2	92,9	27,9	10,0	1,5	100,0	20,0	100,0	4,0	57,6	40	70
4,8	N°. 4	1,6	0,6	2,4	0,7	60,9	9,1	99,9	20,0	100,0	4,0	34,4	20	40
2,0	N°.10	1,1	0,4	1,1	0,3	3,4	0,5	88,1	17,6	99,2	4,0	22,8	10	25
0,075	N°.200	0,5	0,2	0,5	0,2	2,0	0,3	18,5	3,7	0,2	0,0	4,3	0	8

Source: the author.

Figure 24 - Particle size composition of the materials with the addition of 4% ground rubber.

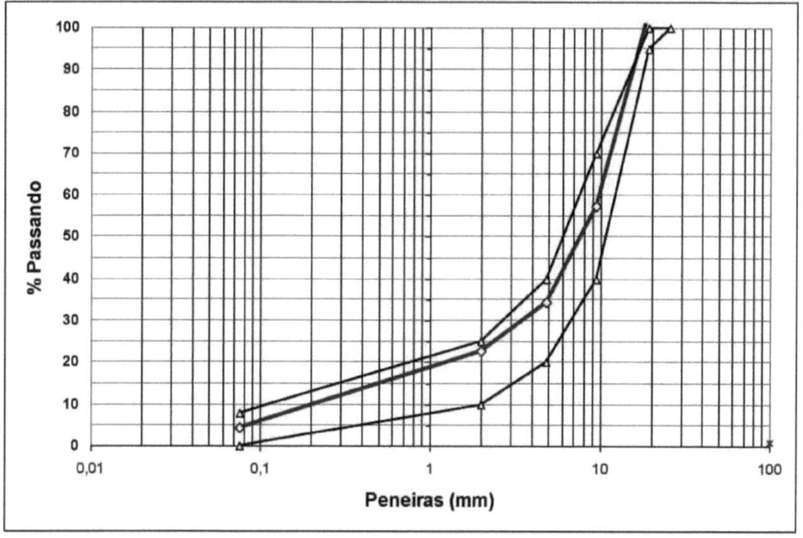

Source: the author.

4.3.4 Dash with 5 % ground rubber added

Table 18 and Figure 25 show the particle size composition of the materials used in the mix with the addition of 5 % ground rubber.

According to the results shown in Table 18, the composite materials for this trait are:

- 35 % 16 mm gravel;
- 30 % 12 mm gravel;
- 15% pebbles;
- 20 % stone dust; e,
- 5 % ground rubber.

The sum of the percentages of these materials falls within band C (gradation) of the DNER-ES 317 (1997) standard - Paving - cold pre-mix.

Figure 25 shows a graph of range C (working range), limited by the black lines and the blue line shows the sum of the percentages of the mix with the addition of 5% ground rubber used in this study.

Table 18 - Particle size composition of the materials with the addition of 5% ground rubber.

Sieves		Material										Grade/Band: C		
		16mm gravel		12mm gravel		Gravel		Stone dust		Rubber		Σ %	Specific Limits	
													cac	the
mm	Pol.	100%	35%	100%	30%	100%	15%	100%	20%	100%	5%	100%	Min.	Max.
25,4	1	100,0	35,0	100,0	30,0	100,0	15,0	100,0	20,0	100,0	5,0	105,0	100	100
19,1	3/4	100,0	35,0	100,0	30,0	100,0	15,0	100,0	20,0	100,0	5,0	105,0	95	100
9,5	3/8	12,0	4,2	92,9	27,9	10,0	1,5	100,0	20,0	100,0	5,0	58,6	40	70
4,8	N°. 4	1,6	0,6	2,4	0,7	60,9	9,1	99,9	20,0	100,0	5,0	35,4	20	40
2,0	N°.10	1,1	0,4	1,1	0,3	3,4	0,5	88,1	17,6	99,2	5,0	23,8	10	25
0,075	N°.200	0,5	0,2	0,5	0,2	2,0	0,3	18,5	3,7	0,2	0,0	4,3	0	8

Source: the author.

Figure 25 - Particle size composition of the materials with the addition of 5% ground rubber.

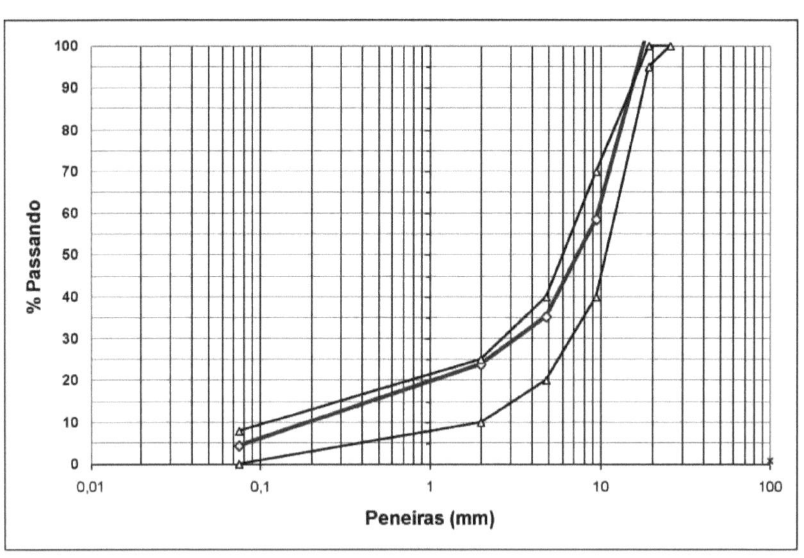

Source: the author.

4.4 Cold asphalt mix

A total of twenty specimens were molded and given the following identifications:

- Test specimens numbers: 1, 2, 3, 4 and 5 - Reference trace (figure 26);
- Test specimens numbers: 7, 8, 9, 10 and 11 - Design with 3% rubber (figure 27);
- Test specimens numbers: 12, 13, 14, 15 and 16 - Design with 4% rubber addition (figure 28); e,
- Test specimens numbers: 17, 18, 19, 20 and 21 - Design with 5% rubber added (figure 29).

Figure 26 - Identification of the molded specimens (1 to 5) of the reference mix.

Source: the author.

Figure 27 - Identification of the molded specimens (7 to 11) of the mix with the addition of 3% rubber.

Source: the author.

Figure 28 - Identification of the molded specimens (12 to 16) of the mix with 4% rubber.

Source: the author.

Figure 29 - Identification of the molded specimens (17 to 21) of the mix with 5% rubber.

Source: the author.

4.4.1 Apparent density

The bulk density test was carried out in accordance with DNER standard ES 317 (1997).

To carry out the test, it is necessary to apply fluidized paraffin to the test specimens, as shown in figure 30.

Figure 30 - Application of fluidized paraffin to the test specimens.

Source: the author.

4.4.1.1 Reference line

Table 19 and figure 31 show the apparent density results of the test specimens of the reference mix (without the addition of ground rubber).

According to the DNER ES 317 standard (1997), the amount of emulsion content for track C should be 4 to 6 %.

Apparent density was determined:

- The immersed weight (g), where the paraffined specimen is immersed in water at room temperature;
- According to DNER-ME 117 (1994), the density of the tape adopted is 0.97 g/cm^3; e,
- According to DNER-ME 117 (1994), the paraffin density adopted is 0.89

g/cm^3.

The apparent density results of the test specimens (01 to 05) are highlighted in the last row of table 19, i.e. 2.173 g/cm^3, 2.157 g/cm^3, 2.142 g/cm^3, 2.127 g/cm$^{(3)}$and 2.115 g/cm$^{(3)}$.

Table 19 - Apparent density of the test specimens of the reference mix.

CP no.	01	02	03	04	05
Emulsion content without water (%)	4,0	4,5	5,0	5,5	6,0
Dimensions (mm)	72,2	74,6	74,8	75,3	73,8
	72,1	74,6	75,0	75,7	73,4
	72,7	74,6	74,9	76,2	72,8
	73,2	74,6	74,6	75,3	72,8
Average Dimension	72,6	74,6	74,8	75,6	73,2
Weight in air (g)	1245,4	1258,6	1270,5	1275,6	1248,2
Air weight+ tape (g)	1248,6	1261,8	1273,1	1278,1	1251,1
Weight in air+tape+paraffin (g)	1255,9	1265,9	1281,9	1286,5	1262,7
Immersed weight (g)	671,2	674,6	676,2	674,8	656,5
Density Tape (g/cm3)	0,97	0,97	0,97	0,97	0,97
Density Paraffin (g/cm3)	0,89	0,89	0,89	0,89	0,89
Apparent Density (g/cm3)	2,173	2,157	2,142	2,127	2,115

Source: the author.

Figure 31 - Graph of the apparent density of the test specimens of the reference mix.

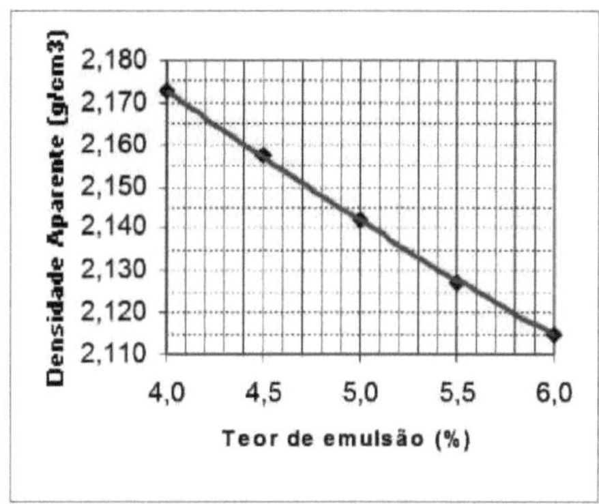

Source: the author.

4.4.1.2 Mix with 3 % ground rubber added

Table 20 and figure 32 show the apparent density results of the test specimens with the addition of 3% ground rubber.

According to the DNER ES 317 standard (1997), the amount of emulsion content for track C should be 4 to 6 %.

The apparent density results of the test specimens (07 to 11) are shown in Table 20, as follows: 2.091 g/cm^3, 2.075 g/cm$^{(3)}$, 2.061 g/cm^3, 2.049 g/cm^3 and 2.036 g/cm$^{(3)}$.

Table 20 - Apparent density of test specimens with 3% ground rubber added.

CP no.	07	08	09	10	11
Emulsion content without water (%)	4,0	4,5	5,0	5,5	6,0
Dimensions (mm)	65,1	65,9	65,4	66,8	65,9
	64,7	65,4	65,3	67,4	66,7
	64,9	65,0	64,7	66,7	66,5
	65,0	65,7	65,2	65,5	66,2
Average Dimension	64,9	65,5	65,2	66,6	66,3
Weight in air (g)	1064,2	1070,6	1076,0	1081,4	1086,1
Air weight+ tape (g)	1067,3	1073,0	1078,7	1084,0	1089,0

Weight in air+tape+paraffin (g)	1072,7	1078,9	1090,4	1089,6	1094,3
Immersed weight (g)	554,5	553,8	552,3	552,8	551,9
Density Tape (g/cm3)	0,97	0,97	0,97	0,97	0,97
Density Paraffin (g/cm3)	0,89	0,89	0,89	0,89	0,89
Apparent Density (g/cm3)	2,091	2,075	2,061	2,049	2,036

Source: the author.

Figure 32 - Apparent density of test specimens with 3% ground rubber added.

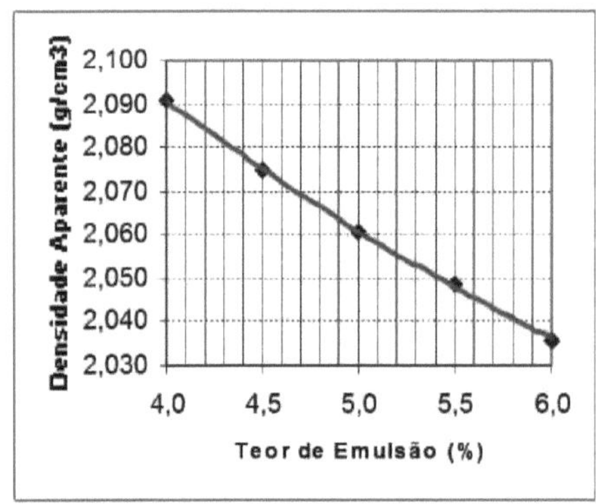

Source: the author.

4.4.1.3 Layout with 4% ground rubber added

Table 21 and Figure 33 show the apparent density results of the test specimens with the addition of 4% ground rubber.

According to the DNER ES 317 standard (1997), the amount of emulsion content for track C should be 4 to 6 %.

The apparent density results of the test specimens (12 to 16) are shown in table 21, i.e. 1.983 g/cm^3, 1.968 g/cm$^{(3)}$, 1.953 g/cm^3, 1.938 g/cm^3 and 1.923 g/cm^3.

Table 21 - Apparent density of test specimens with 4% ground rubber added.

CP no.	12	13	14	15	16
Emulsion content without water (%)	4,0	4,5	5,0	5,5	6,0

Dimensions (mm)	61,2	60,5	62,3	61,8	61,5
	59,9	60,7	61,9	61,2	61,6
	61,1	61,0	62,2	61,9	62,5
	62,6	60,5	61,0	61,6	61,9
Average Dimension	61,2	60,7	61,9	61,6	61,9
Weight in air (g)	931,2	967,7	969,1	978,7	986,1
Air weight+ tape (g)	934,4	970,0	972,1	981,7	989,1
Weight in air+tape+paraffin (g)	944,8	979,1	979,3	994,9	994,7
Immersed weight (g)	460,3	474,8	472,0	472,1	472,5
Density Tape (g/cm³)	0,97	0,97	0,97	0,97	0,97
Density Paraffin (g/cm³)	0,89	0,89	0,89	0,89	0,89
Apparent Density (g/cm³)	1,983	1,968	1,953	1,938	1,923

Source: the author.

Figure 33 - Apparent density of test specimens with 4% ground rubber added.

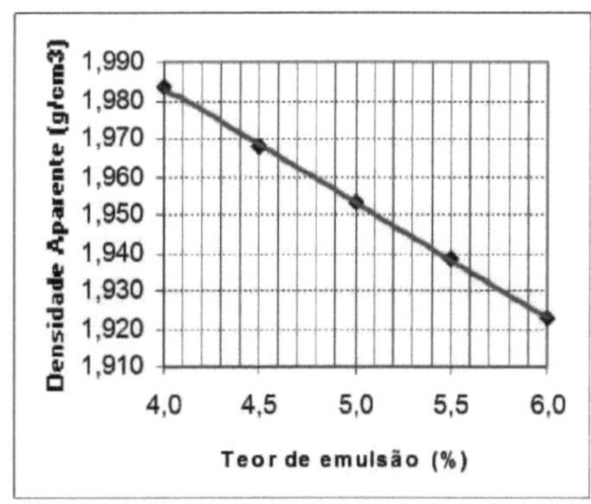

Source: the author.

4.4.1.4 Dash with 5 % ground rubber added

Table 22 and Figure 34 show the values found in the bulk density test of the test specimens with the addition of 5% ground rubber.

According to the DNER ES 317 standard (1997), the amount of emulsion content for track C should be 4 to 6 %.

The results of the apparent density of the test specimens (17 to 21) are respectively: 1.936 g/cm^3, 1.928 g/cm^3, 1.921 g/cm^3, 1.915 g/cm^3 and 1.908 g/cm^3.

Table 22 - Apparent density of test specimens with 5% ground rubber added.

CP no.	**17**	**18**	**19**	**20**	**21**
Emulsion content without water (%)	4,0	4,5	5,0	5,5	6,0
Dimensions (mm)	61,2	63,4	63,5	61,2	62,9
	62,1	63,6	63,5	62,1	63,1
	61,9	62,0	63,3	62,3	63,2
	61,2	63,0	64,1	61,7	63,1
Average Dimension	61,6	63,0	63,6	61,8	63,1
Weight in air (g)	972,2	975,1	981,4	988,4	987,3
Air weight+ tape (g)	974,7	978,0	984,2	991,4	990,6
Weight in air+tape+paraffin (g)	982,1	983,0	993,9	1000,6	996,2
Immersed weight (g)	469,0	468,7	469,2	470,9	469,0
Density Tape (g/cm3)	0,97	0,97	0,97	0,97	0,97
Density Paraffin (g/cm3)	0,89	0,89	0,89	0,89	0,89
Apparent Density (g/cm3)	1,936	1,928	1,921	1,915	1,908

Source: the author.

Figure 34 - Apparent density of test specimens with 5% ground rubber added.

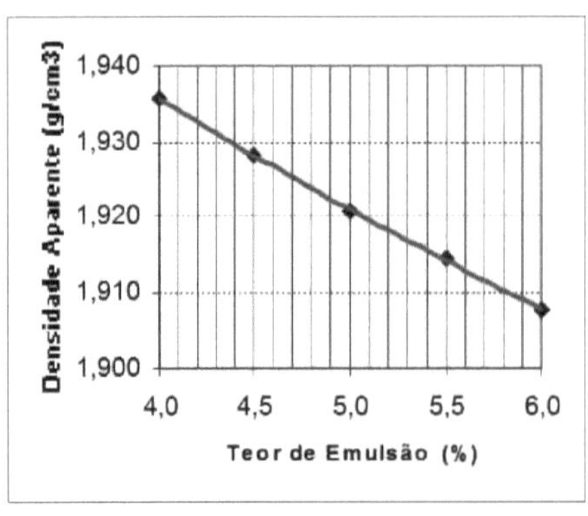

Source: the author.

4.4.2 Marshall stability and fluency

The Marshall stability and creep tests were carried out in accordance with DNER standard ME 107 (1994).

Figures 35, 36, 37, 38 and 39 illustrate, respectively, the addition of the ligant to the aggregate, the addition of the material to be placed in the compaction mold, the mold in the test position on the compaction stand, compaction and the identification of the test specimen.

Figure 35 - ligant added to aggregate.

Source: the author.

Figure 36 - Adding the asphalt mixture to the compaction mold.

Source: the author.

Figure 37 - Compaction mold.

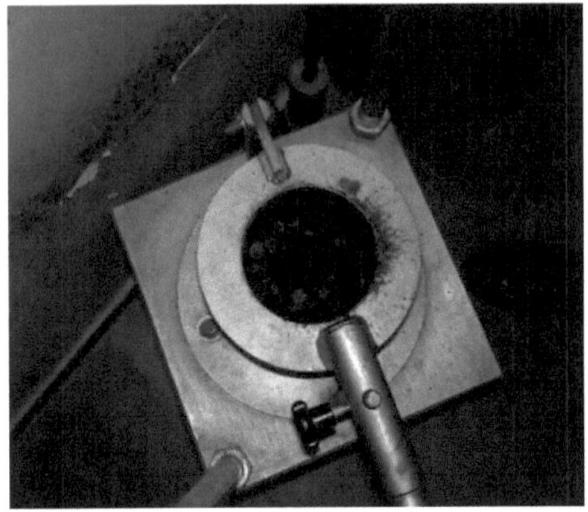

Source: the author.

Figure 38 - Compaction of the test specimen.

Source: the author.

Figure 39 - Identification of the test specimen.

Source: the author.

Figure 40 shows the rupture of the test specimen to obtain the stability and creep results.

Figure 40 - Rupture of the test specimen.

Source: the author.

4.4.2.1 Reference line

The results of the test specimens of the reference mix (without the addition of ground rubber) found for Marshall stability and creep are shown in Table 23 and Figures 41 and 42.

The load in Kgf required to cause the specimen to break is the read stability. This value is corrected for the thickness of the test specimen by multiplying it by a factor that depends on the thickness of the specimen. The result obtained is the corrected stability.

Table 23 - Marshall stability and creep of the test specimens of the reference mix.

CP no.	01	02	03	04	05
Emulsion content without water (%)	4,0	4,5	5,0	5,5	6,0
L.I. Creep (mm)	2,13	2,25	1,42	2,05	0,71
L.F. Creep (mm)	4,68	5,15	4,96	6,14	5,65
Deflectometer reading (div.)	568	428	343	287	235
Read Stability (kgf)	950	716	574	480	393
Correction Factor	0,82	0,78	0,78	0,77	0,81
Creep (mm)	2,6	2,9	3,5	4,1	4,9
Corrected Stability (kgf)	779	559	448	370	318

Source: the author.

Figure 41 - Marshall stability of the test specimens of the reference mix.

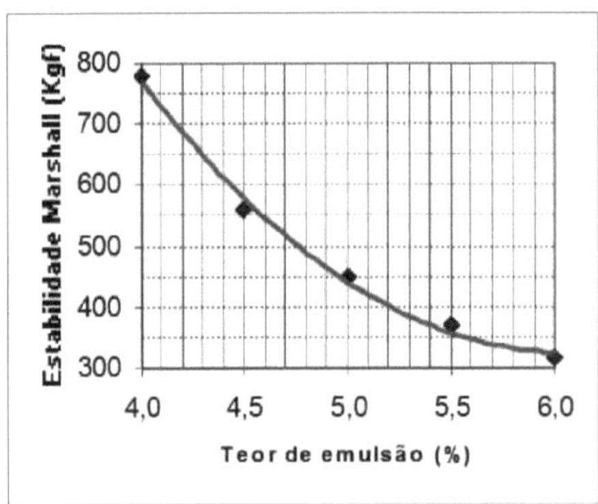

Source: the author.

Table 23 and figure 42 show the Marshall stability results of the test specimens (1 to 5), respectively:

- For an emulsion content of 4.0 %, the Marshall stability is 779 kgf;

- For an emulsion content of 4.5%, the Marshall stability is 559 kgf;
- For an emulsion content of 5.0 %, the Marshall stability is 448 kgf;
- For an emulsion content of 5.5%, the Marshall stability is 370 kgf; e,
- For an emulsion content of 6.0 %, the Marshall stability is 318 kgf.

Figure 12 - Creep of the test specimens of the reference mix.

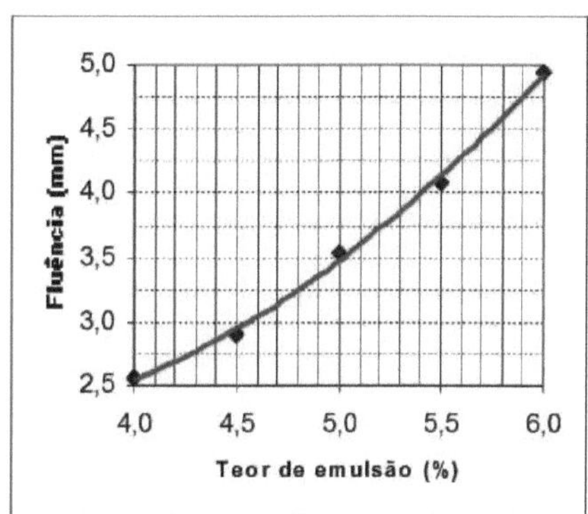

Source: the author.

Table 23 and figure 43 show the results of the creep test:

- To ocontentcontentemulsion4,0% a creep é of 2,6 mm;
- To ocontentcontentemulsion4,5% a creep é of 2,9 mm;
- To ocontentcontentemulsion5,0% a creep é of 3,5 mm;
- For an emulsion content of 5.5%, the creep is 4.1 mm; e,
- For an emulsion content of 6.0 %, the creep is 4.9 mm.

4.4.2.2 Layout with 3 % ground rubber added

Table 24 and Figures 44 and 45 show the results of the Marshall stability and creep of the test specimens with the addition of 3% ground rubber.

Table 24 - Marshall stability and creep of test specimens with the addition of 3% ground rubber.

CP no.	07	08	09	10	11

Emulsion content without water (%)	4,0	4,5	5,0	5,5	6,0
L.I. Creep (mm)	3,01	2,12	2,49	2,63	2,53
L.F. Creep (mm)	7,40	7,13	7,93	8,63	8,88
Deflectometer reading (div.)	277	247	225	185	140
Read Stability (kgf)	463	413	376	310	234
Correction Factor	0,96	0,95	0,96	0,93	0,94
Creep (mm)	4,4	5,0	5,4	6,0	6,4
Corrected Stability (kgf)	445	393	361	288	220

Source: the author.

Figura 13 - Marshall stability of test specimens with the addition of 3% ground rubber.

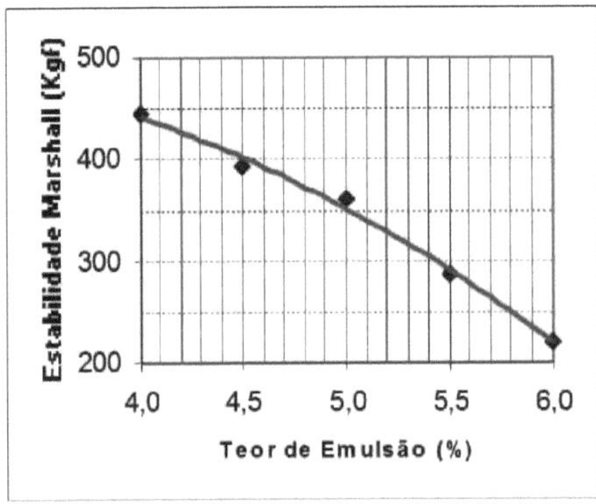

Source: the author.

Table 24 and Figure 44 show the Marshall stability results of the test specimens (07 to 11):

- For an emulsion content of 4.0 %, the Marshall stability is 445 kgf;
- For an emulsion content of 4.5%, the Marshall stability is 393 kgf;
- For an emulsion content of 5.0 %, the Marshall stability is 361 kgf;
- For an emulsion content of 5.5%, the Marshall stability is 288 kgf; e,
- For an emulsion content of 6.0 %, Marshall stability is 220 kgf.

Figura 14 - Creep of test specimens with the addition of 3% ground rubber.

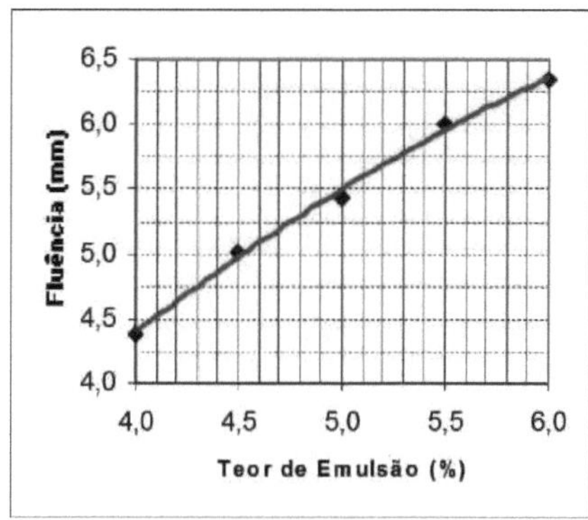

Source: the author.

Table 24 and figure 45 show the results of the creep test:

- To oemulsionemulsion 4,0 % a creep éof4,4mm;
- To oemulsionemulsion 4,5 % a creep éof5,0mm;
- To oemulsionemulsion 5,0 % a creep éof5,4mm;
- To oemulsionemulsion 5,5 % a creep éof6,0mm; e,
- To oemulsionemulsion 6,0 % a creep éof6,4mm.

4.4.2.3 Layout with 4 % ground rubber added

Table 25 and Figures 46 and 47 show the results of the Marshall stability and creep of the test specimens with the addition of 4% ground rubber.

Table 25 - Marshall stability and creep of test specimens with the addition of 4% ground rubber.

CP no.	12	13	14	15	16
Emulsion content without water (%)	4,0	4,5	5,0	5,5	6,0
L.I. Creep (mm)	2,03	2,71	2,93	2,74	2,44
L.F. Creep (mm)	6,03	7,41	8,09	8,29	8,62
Deflectometer reading (div.)	213	187	173	162	145
Read Stability (kgf)	356	313	289	271	243
Correction Factor	1,02	1,08	1,04	1,05	1,04

Creep (mm)	4,0	4,7	5,2	5,6	6,2
Corrected Stability (kgf)	363	338	301	285	252

Source: the author.

Figura 15 - Marshall stability of test specimens with the addition of 4% ground rubber.

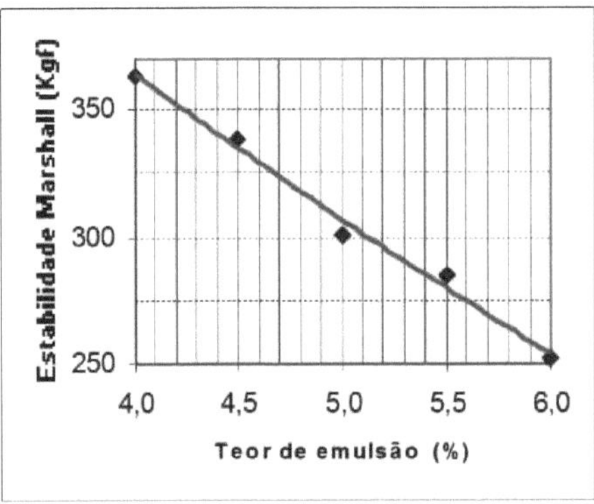

Source: the author.

Table 25 and figure 46 show the results of the Marshall stability of the test specimens (12 to 16), respectively:

- For an emulsion content of 4.0 %, the Marshall stability is 363 kgf;
- For an emulsion content of 4.5%, the Marshall stability is 338 kgf;
- For an emulsion content of 5.0 %, the Marshall stability is 301 kgf;
- For an emulsion content of 5.5%, the Marshall stability is 285 kgf; e,
- For an emulsion content of 6.0 %, the Marshall stability is 252 kgf.

Figura 16 - Creep of test specimens with the addition of 4% ground rubber rubber.

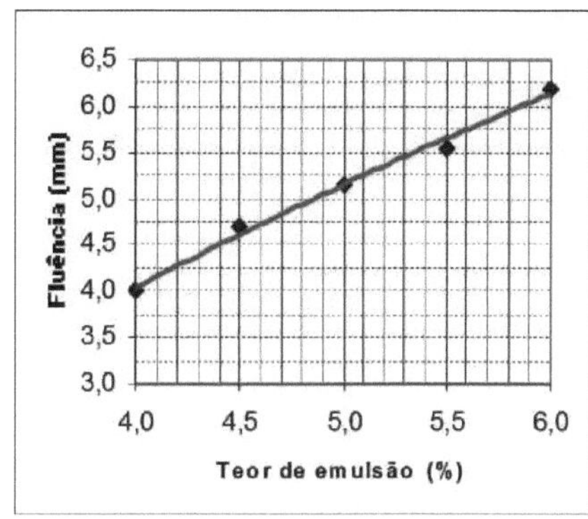

Source: the author.

Table 25 and Figure 47 show the results of the creep test:

- To ocontentofemulsion content4,0 % thecreep is4,0 mm;
- To ocontentofemulsion content4,5 % thecreep is4,7 mm;
- To ocontentofemulsion content5,0 % thecreep is5,2 mm;
- To ocontentofemulsion content5,5 % thecreep is5,6 mm; and,
- To ocontentof6,0 content% thecreep is6,2 mm.

4.4.2.4 Layout with 5 % ground rubber added

Table 26 and Figures 48 and 49 show the results of the Marshall stability and creep of the test specimens with the addition of 5% ground rubber.

Table 26 - Marshall stability and creep of test specimens with the addition of 5% ground rubber.

CP no.	17	18	19	20	21
Emulsion content without water (%)	4,0	4,5	5,0	5,5	6,0
L.I. Creep (mm)	2,34	2,54	3,03	2,54	3,17
L.F. Creep (mm)	7,42	8,39	9,68	10,09	11,04
Deflectometer reading (div.)	181	175	161	138	106
Read Stability (kgf)	303	293	269	231	177

Correction Factor	1,05	1,01	1,00	1,04	1,01
Creep (mm)	5,1	5,9	6,7	7,6	7,9
Corrected Stability (kgf)	318	296	269	240	179

Source: the author.

Figure 17 - Marshall stability of the test specimens with the addition of 5% ground rubber.

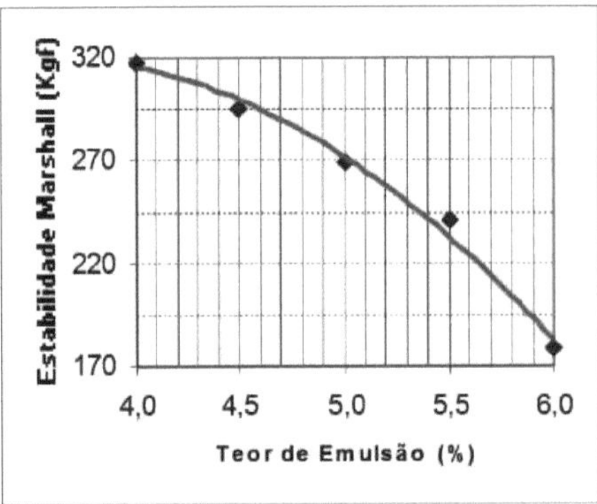

Source: the author.

Table 26 and figure 48 show the Marshall stability results of the test specimens (17 to 21) respectively:

- For an emulsion content of 4.0 %, the Marshall stability is 318 kgf;
- For an emulsion content of 4.5%, the Marshall stability is 296 kgf;
- For an emulsion content of 5.0 %, the Marshall stability is 269 kgf;
- For an emulsion content of 5.5%, the Marshall stability is 240 kgf; e,
- For an emulsion content of 6.0 %, the Marshall stability is 179 kgf.

Figure 18 - Creep of the test specimens with the addition of 5% ground rubber.

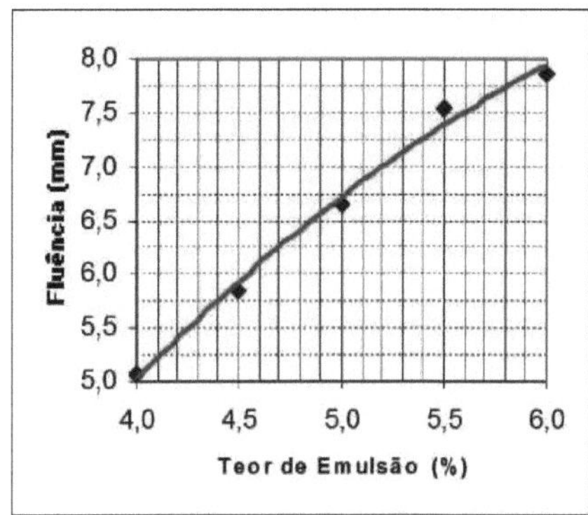

Source: the author.

Table 26 and figure 49 show the results of the creep test:

- To oemulsion emulsion 4,0 % a creep é of 5,1mm;
- To oemulsion emulsion 4,5 % a creep é of 5,9mm;
- To oemulsion emulsion 5,0 % a creep é of 6,7mm;
- To oemulsion emulsion 5,5 % a creep é of 7,6mm; e,
- To oemulsion emulsion 6,0 % a creep é of 7,9mm.

4.4.3 Mixture characteristics

Sub-items 5.4.3.1, .4.3.2, 5.4.3.3 and 5.4.3.4 present the characteristics of the mixtures of the traits studied: reference, with additions of 3%, 4% and 5% ground rubber.

4.4.3.1 Reference line

Table 27 and Figures 50 and 51 show the individual results of the average density of the mixture, the mineral aggregate voids, the volume filled with bitumen, the theoretical density, the volume of voids and the bitumen voids ratio of the test specimens of the reference mix (without the addition of ground rubber).

Table 27 - Characteristics of the mixture of test specimens of the reference mix.

CP no.	01	02	03	04	05

Average Density of the Mixture (g/cm3)	2,846				
Mineral aggregate voids (M.A.V.) (%)	26,7	27,6	28,5	29,4	30,1
Volume Full Bitumen (V.C.B.) (%)	8,6	9,6	10,6	11,6	12,6
Theoretical Density (g/cm3)	2,653	2,631	2,609	2,588	2,567
Empty volume (%)	18,1	18,0	17,9	17,8	17,6
Bitumen to voids ratio (B.V.R.) (%)	32,2	34,8	37,2	39,4	41,6

Source: the author.

Figura 19 - Void index of the test specimens of the reference mix.

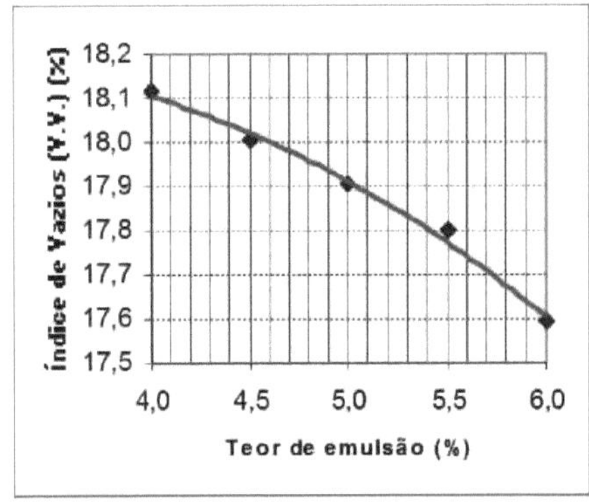

Source: the author.

Figura 20 - Bitumen to voids ratio of the test specimens of the reference mix.

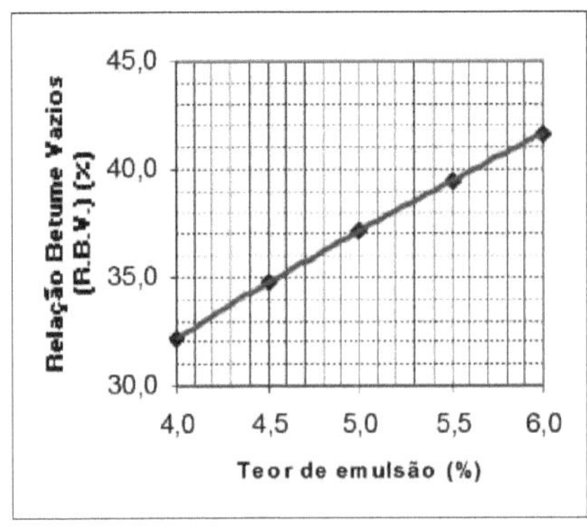

Source: the author.

Table 28 shows the values specified by and the values obtained in the tests of the reference trace.

Table 28 - Values specified by the standard and values obtained in the reference trace tests.

Specified Values (DNER ME 107, 1994)	
Emptiness Index (V.V.) (%)	5 30
Creep (mm)	2 4,5
Stability at 40°C (kgf)	Minimum: 250
Values obtained in the test	
Emptiness Index (V.V.) (%)	17,9
Creep (mm)	3,5
Stability at 40°C (kgf)	448
Apparent Density ($g/cm3$)	2,142
Bitumen to voids ratio (B.V.R.) (%)	37,2
Optimum Emulsion Content (%)	5,0

Source: DNER ME 107, 1994; the author

According to the results shown in Table 28 of the reference mix tests, all the values meet the specifications of DNER standard 107 (1994).

In accordance with DNER standard 107 (1994), the following applies:

- To determine the optimum emulsion content, the average of the contents is calculated and is equal to 5.0 %;
- To determine the void ratio, the average of the range (values specified by the standard) is calculated and is equal to 17.5%;
- For the determination of creep, the average of the range (values specified by standard) is calculated, being equal to 3.25 %; e,
- For stability, the minimum value in compliance with the standard is 250 kgf.

4.4.3.2 Layout with 3 % ground rubber added

Table 29 and Figures 52 and 53 show the individual results of the average density of the mixture, the mineral aggregate voids, the volume filled with bitumen, the theoretical density, the volume of voids and the bitumen to voids ratio of the test specimens with the addition of 3% ground rubber.

Table 29 - Characteristics of the mixture of test specimens with the addition of 3% ground rubber.

CP no.	07	08	09	10	11
Average Density of the Mixture (g/cm3)	2,822				
Mineral aggregate voids (M.A.V.) (%)	28,9	29,8	30,6	31,4	32,2
Volume Full Bitumen (V.C.B.) (%)	8,3	9,2	10,2	11,1	12,1
Theoretical Density (g/cm3)	2,633	2,611	2,590	2,569	2,548
Empty volume (%)	20,6	20,6	20,4	20,2	20,1
Bitumen to voids ratio (B.V.R.) (%)	28,7	31,0	33,3	35,5	37,5

Source: the author.

Figura 21 - Void index of test specimens with the addition of 3% ground rubber.

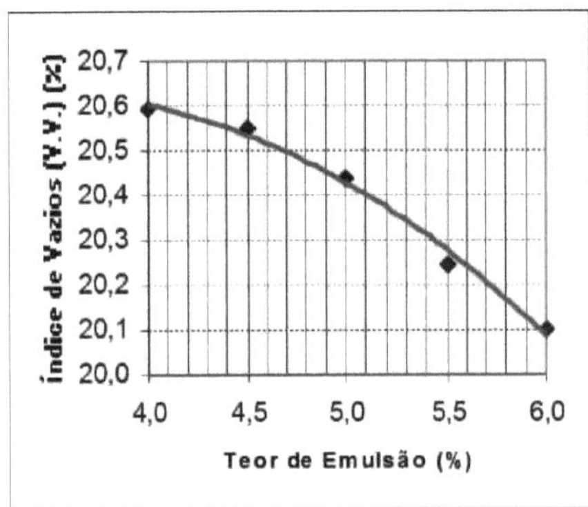

Source: the author.

Figura 22 - Bitumen to voids ratio of test specimens with the addition of 3% ground rubber.

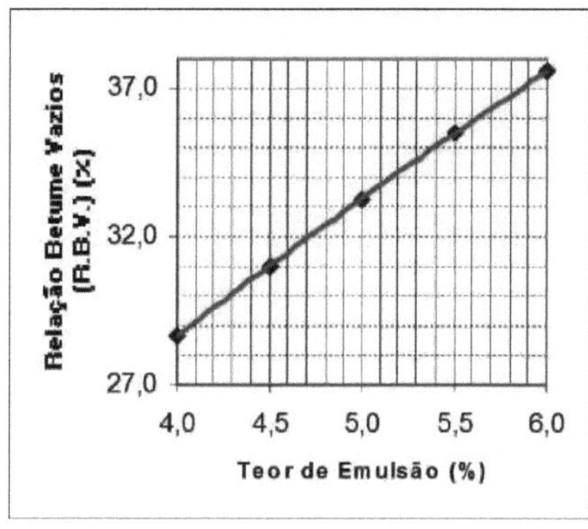

Source: the author.

Table 30 shows the test results obtained from the mix with the addition of 3% ground rubber.

Table 30 - Values obtained in the tests of the mix with the addition of 3% ground rubber.

Values obtained in the test	
Emptiness Index (V.V.) (%)	20,6
Creep (mm)	4,4
Stability at 40°C (kgf)	445
Apparent Density (g/cm^3)	2,091
Bitumen to voids ratio (B.V.R.) (%)	28,7
Optimum Emulsion Content (%)	4,0

Source: the author.

Table 30 shows that all the test results for the mix with the addition of 3% ground rubber meet the specifications of DNER standard 107 (1994).

4.4.3.3 Layout with 4 % ground rubber added

Table 31 and Figures 54 and 55 show the individual results obtained from the tests on the average density of the mix, the voids in the mineral aggregate, the volume filled with bitumen, the theoretical density, the volume of voids and the bitumen to voids ratio of the test specimens for the mix with the addition of 4% ground rubber.

Table 31 - Characteristics of the mixture of test specimens the addition of 4% ground rubber.

CP no.	12	13	14	15	16
Average Density of the Mixture (g/cm³)	2,810				
Mineral aggregate voids (M.A.V.) (%)	32,2	33,1	34,0	34,8	35,7
Volume Full Bitumen (V.C.B.) (%)	7,8	8,8	9,7	10,5	11,4
Theoretical Density (g/cm³)	2,623	2,602	2,580	2,560	2,539
Empty volume (%)	24,4	24,4	24,3	24,3	24,3
Bitumen to voids ratio (B.V.R.) (%)	24,3	26,5	28,4	30,3	32,0

Source: the author.

Figura 23 - Void index of the test specimens with the addition of 4% ground rubber.

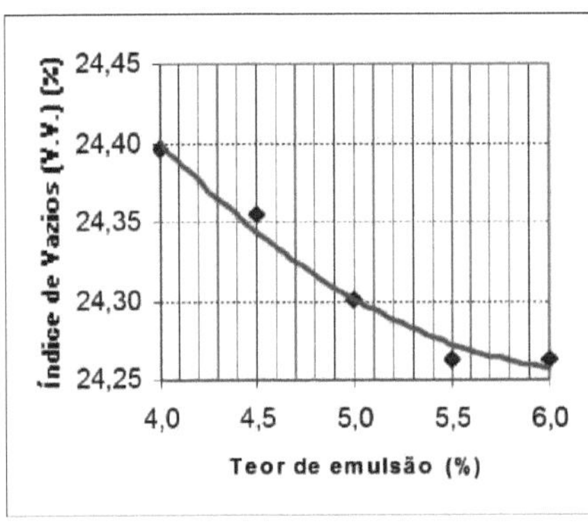

Source: the author.

Figura 24 - Bitumen to voids ratio of the test specimens with the addition of 4% ground rubber.

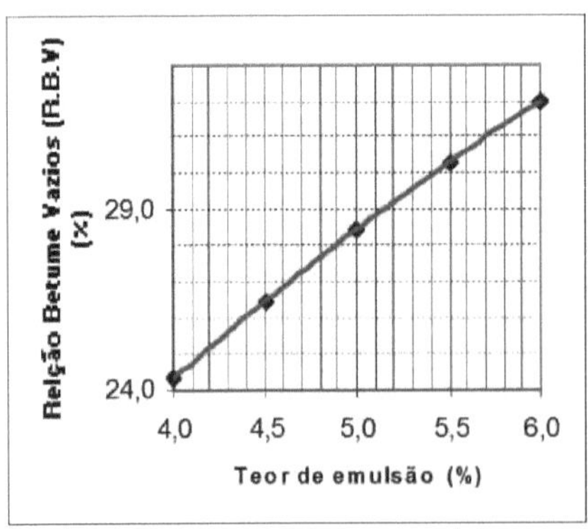

Source: the author.

Table 32 below shows the test results for the mix with 4 % ground rubber.

Table 32 - Values obtained in the tests of the mix with the addition of 4 % ground rubber.

Values obtained in the test	
Emptiness Index (V.V.) (%)	24,3
Creep (mm)	4,7
Stability at 40°C (kgf)	338
Apparent Density (g/cm³)	1,968
Bitumen to voids ratio (B.V.R.) (%)	26,5
Optimum Emulsion Content (%)	4,5

Source: the author.

It can be seen that the results meet the specifications of the DNER 107 (1994) standard, with the exception of the creep test which showed a result of 4.7 mm. This value is higher than that established by the DNER 107 (1994) standard.

4.4.3.4 Layout with 5 % ground rubber added

Table 33 and Figures 56 and 57 show the results of the characteristics of the mixture with the addition of 5% ground rubber, with regard to the following tests: average density of the mixture, mineral aggregate voids, volume filled with bitumen, theoretical density, volume of voids and bitumen voids ratio.

Table 33 - Characteristics of the mixture of test specimens with the addition of 5% ground rubber.

CP no.	17	18	19	20	21
Average Density of the Mixture (g/cm³)	2,805				
Mineral aggregate voids (M.A.V.) (%)	33,7	34,4	34,9	35,5	36,1
Volume Full Bitumen (V.C.B.) (%)	7,7	8,6	9,5	10,4	11,3
Theoretical Density (g/cm³)	2,619	2,598	2,576	2,556	2,535
Empty volume (%)	26,1	25,8	25,4	25,1	24,7
Bitumen to voids ratio (B.V.R.) (%)	22,7	25,0	27,2	29,3	31,4

Source: the author.

Figura 25 - Void index of test specimens with the addition of 5% ground rubber.

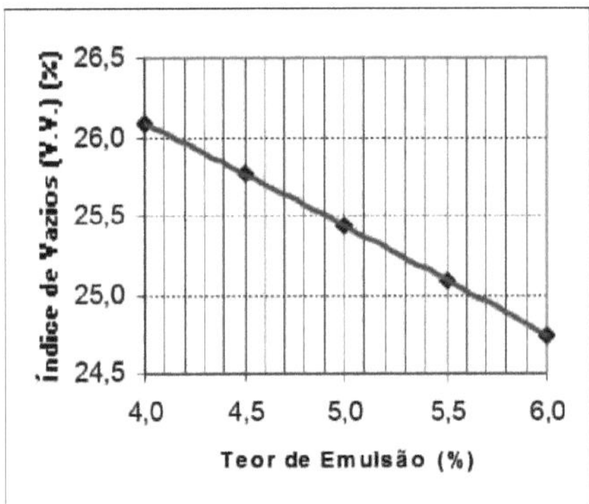

Source: the author.

Figura 26 - Bitumen to voids ratio of test specimens with the addition of 5% ground rubber.

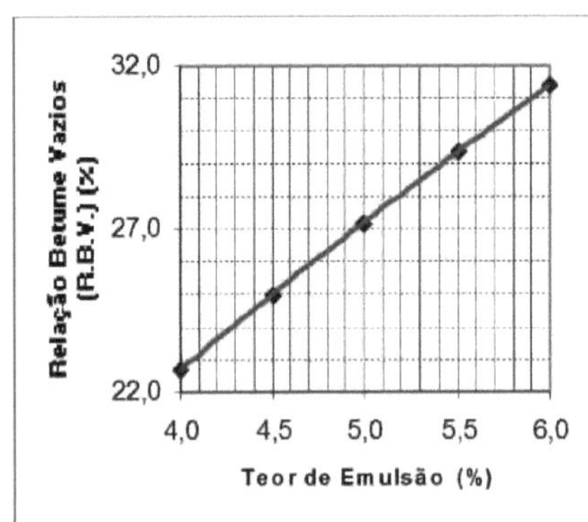

Source: the author.

Table 34 shows the final test results for the mix with 5 % ground rubber.

Table 34 - Values obtained in the tests with the addition of 5% ground rubber.

Values obtained in the test	
Emptiness Index (V.V.) (%)	26,0
Creep (mm)	5,0
Stability at 40°C (kgf)	320
Apparent Density (g/cm3)	1,936

Bitumen to voids ratio (B.V.R.) (%)	22,7
Optimum Emulsion Content (%)	4,0

Source: the author.

Table 34 shows that all the results meet the specifications of DNER standard 107 (1994), with the exception of the creep test, which showed a result higher than the value established by the standard (range 2 to 4.5%).

4.5 Comparison of results

Comparisons of the test results for all the traits studied are presented below: reference, 3 %, 4 % and 5 % ground rubber additions.

4.5.1 void ratio

Figure 58 shows the results obtained from the void ratio test in relation to the reference mix (without the addition of ground rubber) and it can be seen that the best mix with the lowest void ratio is the mix with the addition of 3% ground rubber.

Figure 27 - Emptiness index test results according to DNER standard 107 (1994).

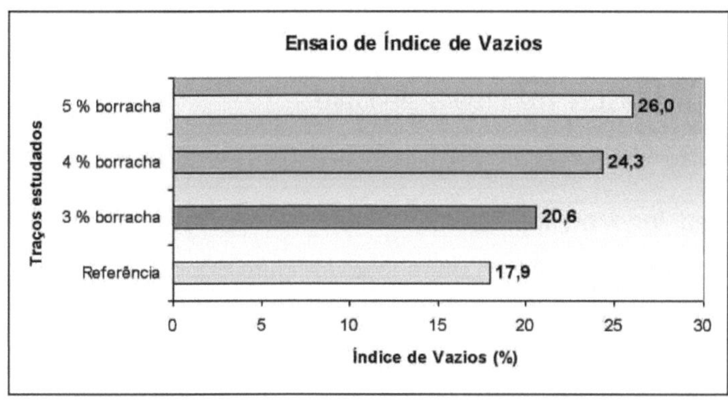

Source: the author.

According to figure 58, all the results meet the specifications of DNER standard 107 (1994), whose values should be between 5 and 30% (average 17.5%).

4.5.2 Fluency

Comparing the results obtained from the creep test in relation to the reference mix (without the addition of ground rubber), it can be seen that the best mix with the lowest creep is the mix with the addition of 3% ground rubber, as can be seen in figure 59.

Figure 28 - Creep test results according to DNER standard 107 (1994).

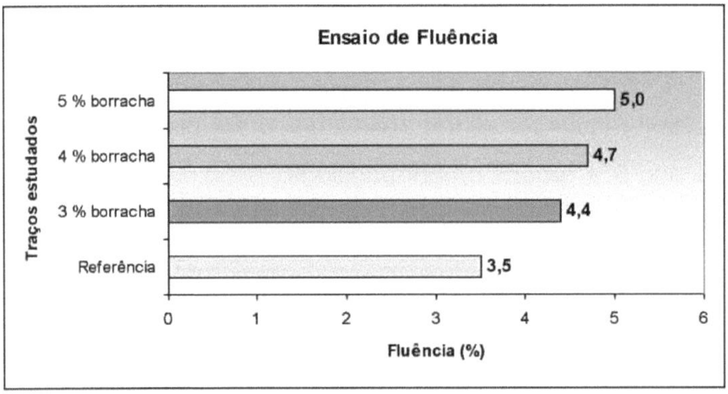

Source: the author.

It can be seen that the mix with the addition of 3% rubber meets the specifications of DNER standard 107 (1994), whose values should be between 2 and 4.5% (average 3.25%). The mixtures with 4 and 5% rubber do not meet the specifications of the DNER 107 (1994) standard.

4.5.3 Marshall stability

The results obtained for Marshall stability in relation to the reference mix (without the addition of ground rubber) are shown in figure 60, where it can be seen that the best mix that meets the highest Marshall stability is the mix with the addition of 3% ground rubber.

Figure 29 - Marshall stability test results according to DNER standard 107 (1994).

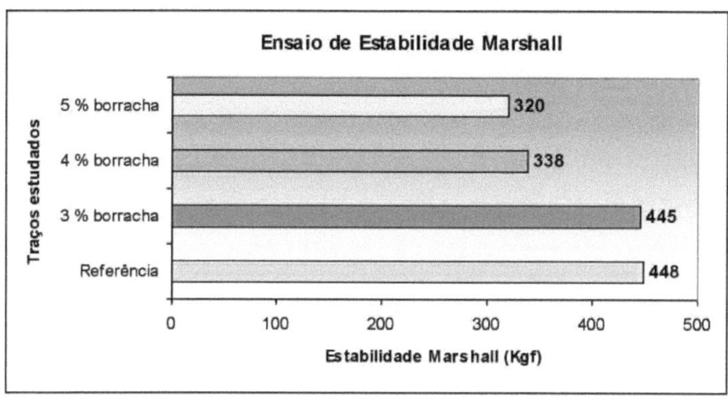

Source: the author.

4.5.4 Apparent density

Comparing the results obtained (figure 61) for apparent density in relation to the reference mix (without the addition of ground rubber), it can be seen that the higher the percentage of added rubber, the lower the apparent density.

Figure 30 - Apparent density test results according to DNER standard 107 (1994).

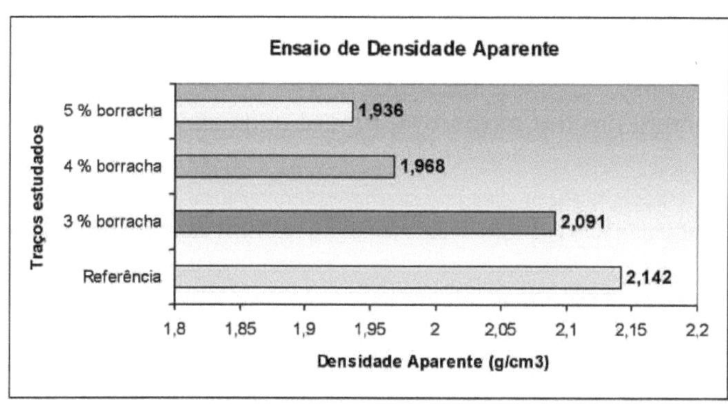

Source: the author.

It can be seen that the best mix studied is the one with the addition of 3% rubber, as this is the mix with the highest apparent density.

4.5.5 Bitumen to voids ratio (B.V.R.)

A comparative study of the bitumen to voids ratio test can be seen in figure 62, in relation to the reference mix (without the addition of ground rubber), where it can be seen that the best mix that meets the highest bitumen to voids ratio is the mix with the addition of 3% ground rubber.

Figure 31 - Results of the bitumen-voids ratio tests according to DNER standard 107 (1994).

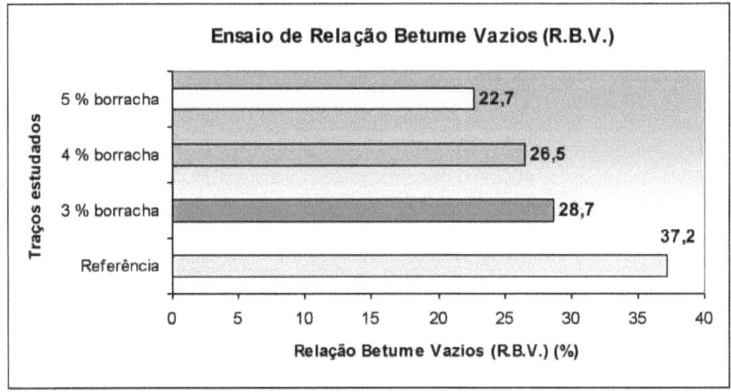

Source: the author.

4.5.6 Optimum emulsion content

Figure 63 shows a comparison of the results obtained for the optimum emulsion content in relation to the reference mix (without the addition of ground rubber). It can be seen that the best mixes that meet the lowest amount of optimum emulsion content are the mixes with rubber additions of 3 and 5 % ground rubber.

Figure 32 - Results of the bitumen to voids ratio tests according to DNER standard 107 (1994).

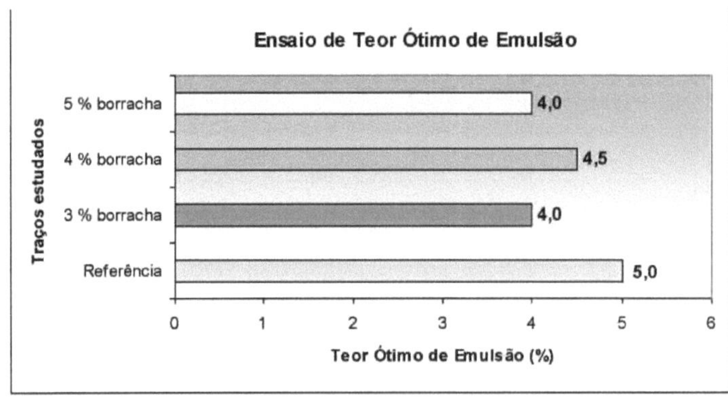

Source: the author.

5 CONCLUSIONS

The incorporation of ground tire rubber into cold asphalt mixtures proved to be valid. The high consumption of tires in these mixtures proves the effectiveness of using them to reduce the number of old tires. With this, the use of rubber from tires in asphalt paving presents an interesting and environmentally correct solution to the problem of this waste.

Comparing all the results of the asphalt mixtures in relation to the reference mix, we can conclude:

- The mix with the addition of ground rubber had: the lowest void ratio, the lowest creep, the highest Marshall stability, the highest apparent density, the highest bitumen to voids ratio and the lowest asphalt emulsion content;
- Trait with the addition of 4% ground rubber: in the creep test it showed a result above that specified in DNER standard 107 (1994), not meeting the conditions of the standards in force, and therefore discarded; e,
- The mix with the addition of 5% ground rubber had the lowest emulsion content, but in the creep test it showed a result above that specified in DNER standard 107 (1994), not meeting the conditions of the standards in force and therefore discarded.

Among the mixes studied, the mix with the addition of 3% rubber is the best mix, as it guarantees:

- That the asphalt does not oxidize due to the action of water and/or air and that there is no leaking, as it has the smallest volume of voids;
- Maximum stability because it has the highest bulk density;
- The highest resistance to diametrical compression and the greatest Marshall stability;
- The lowest consumption of asphalt emulsion as it has the lowest optimum content of asphalt emulsion; e,

- That there is enough bitumen and that it does not leach out because it has the highest bitumen to voids ratio.

REFERENCES

ABEDA, Brazilian Association of Asphalt Distributors. **Basic Manual of Asphalt Emulsions: Solutions for Paving Your City**. Brazil, 2001.

AIKMAN, E. Engineering Brief No 23- **Open-Graded Asphalt Emulsion Mixes**. Available at: http://www.faa.gov. Accessed on: 25 Jan. 2006.

AIREY, G.D.; RAHMAN, M.; COLLOP, **A The Influence of Crude Source and Penetration Grade on the Interaction of Crumb Rubber and Bitumen**. Proceedings Asphalt Rubber 2003. Brazil, 2003.

ANIP, National Association of the Tyre Industry. **Sector Activity Indicators**. Available at: http://www.anip.com.br Accessed on: September 10, 2007.

ASSOCIAÇAO BRASILEIRA DE NORMAS TÉCNICAS NRB 14491: **Emulsões asfàlticas - determinação da viscosidade Saybolt Furol.** Rio de Janeiro: ABNT, April 2007.

ASSOCIAÇAO BRASILEIRA DE NORMAS TECNICAS NRB 6300: **Cationic asphalt emulsions - determination of water resistance (adhesiveness)**. Rio de Janeiro: ABNT, April 2001.

BRAZILIAN ASSOCIATION OF TECHNICAL STANDARDS NRB 6568: **Asphalt emulsions - determination of distillation residue**. Rio de Janeiro: ABNT, September 2005.

ASPHALT INSTITUTE **The asphalt handbook Manual Series** n°. 4 (Ms-4), p. 43. UsA, 1989.

BERTOLLO, S. A. M.; Fernandes Jûnior, J. L.; Villaverde, R. B.; Migotto Filho, D. (2000) **Asphalt paving: an alternative for the reuse of used tires**. Revista de Limpeza Pùblica, ed. 54. pp. 23-30.

BROSSEAUD, Y. **French Asphalt Coatings: Overview of Techniques, Balance of Behavior.** 16th Asphalt Meeting. Brazil, 2002.

CLERMAN, D. S. **Laboratory Study of Cold Asphalt Mixtures Produced**

with the Incorporation of Tire Rubber. Master's thesis. Federal University of Rio Grande do Sul. Brazil, 2004.

NATIONAL ENVIRONMENT COUNCIL - CONAMA Available at <www.mma.gov.br/port/conama> **Resolution 258**, August 26, 1999. Accessed on: September 14, 2007.

CNT - National Transportation Confederation. **CNT Bulletin June 2007**. Available at <http://www.cnt.org.br>. Accessed on: September 25, 2007.

NATIONAL NATIONAL DE ROADS E ROADS - **DNER ME 002: Asphalt Emulsion - Particle Load,** 1998.

NATIONAL NATIONAL DE ROADS E ROADS - **DNER ME 003: Bituminous Material - Determination of Penetration,** 1999.

DEPARTAMENTO NACIONAL DE ESTRADAS E RODAGEM - **DNER ME 005: Emulsâo Asfàltica - Determinaçâo da Peneiraçâo (ABNT - NBR 14393),** 1995.

DEPARTAMENTO NACIONAL DE ESTRADAS E RODAGEM - **DNER ME 006: Asphalt Emulsions - Determination of Sedimentation**, 2000.

DEPARTAMENTO NACIONAL DE ESTRADAS E RODAGEM - **DNER ME 007: Emulsâo Asfàltica - Determinação da Ruptura - Método da Mistura com Cimento (ABNT NBR 6297)**, 1994.

DEPARTAMENTO NACIONAL DE ESTRADAS E RODAGEM - **DNER ME 008: Emulsâo Asfàltica - Determinação da Ruptura - Mètodo da Mistura com Filer Silicico (ABNT NBR 6302)**, 1994.

DEPARTAMENTO NACIONAL DE ESTRADAS E RODAGEM - **DNER ME 010: Petroleum Asphalt Cements - Determination of Bitumen Content (ABNT MB 166)**, 1994.

DEPARTAMENTO NACIONAL DE ESTRADAS E RODAGEM - **DNER ME 035: Aggregates - determination of "Los Angeles" abrasion**, 1998.

NATIONAL NATIONAL DE ROADS E ROADS- **DNER ME 054: Sand equivalent**, 1997.

NATIONAL NATIONAL DE ROADS E ROADS- **DNER ME059: Emulsions asphalt - determination of water resistance (adhesiveness) (ABNT NBR 6300)**, 1994.

NATIONAL NATIONAL DE ROADS E ROADS **-DNER ME 063: Cationic asphalt emulsions - determination of demulsibility**, 1994.

DEPARTAMENTO NACIONAL DE ESTRADAS E RODAGEM - **DNER ME 083: Aggregates - granulometric analysis**, 1994.

DEPARTAMENTO NACIONAL DE ESTRADAS E RODAGEM - **DNER ME 086: Aggregate - determination of shape index**, 1994.

DEPARTAMENTO NACIONAL DE ESTRADAS E RODAGEM - **DNER ME 089: Aggregates - durability assessment using sodium or magnesium sulphate solutions**, 1994.

DEPARTAMENTO NACIONAL DE ESTRADAS E RODAGEM - **DNER ME 107: Cold Bituminous Mix with Asphalt Emulsion - Marshall Test**, 1994.

DEPARTAMENTO NACIONAL DE ESTRADAS E RODAGEM - **DNER ME 117: Bituminous Mixture - Determination of Apparent Density**, 1994.

DEPARTAMENTO NACIONAL DE ESTRADAS E RODAGEM - **DNER ME 149: Cationic asphalt emulsions - pH determination (ABNT NBR 6299)**, 1994.

DEPARTAMENTO NACIONAL DE ESTRADAS E RODAGEM - **DNER ME 163: Bituminous materials - determination of ductility**, 1998.

DEPARTAMENTO NACIONAL DE ESTRADAS E RODAGEM - **DNER PRO 199: Reduction of aggregate field samples for laboratory testing**, 1996

DEPARTAMENTO NACIONAL DE ESTRADAS E RODAGEM - **DNER ES**

317: Pavimentação - Pré-misturados a frio, 1997.

NATIONAL NATIONAL DE ROADS E ROADS - **DNER ES 369: Cationic Asphalt Emulsions**, 1997.

NATIONAL NATIONAL DE ROADS E ROADS - **DNER ES 395: Paving - bonding paint with polymer asphalt,** 1999.

DEPARTAMENTO NACIONAL DE ESTRADAS E RODAGEM - **DNER EM 396: Polymer-modified asphalt cement,** 1999.

NATIONAL DEPARTMENT OF ROADS AND HIGHWAYS - DNER. **Paving Manual**, 1996.

ECHIMENCo, L. **Used tires yield profits**. In: Jornal o Estado de Sao Paulo. Sao Paulo, April 17, 2001. Business Panel, p. pn. 1, 3 e 4.

FERNANDEZ DEL CAMPO, J. A. **Tratado de estabilización y reciclado de capas de firmes con emulsión asfàltica.** Spain, 1998.

GOODYEAR. Charles Goodyear and the Strange Story of Rubber. 1999. Available at< http://www.goodyear.com/corporate_information>. Accessed on Feb. 15, 1999.

GRECA, M.R.; MORILHA, A.J. **Asfalto Borracha** - Ecoflex, Curitiba, 2003.

HEITZMAN, M. **Design and Construction of Asphalt Paving Materials with Crumb Rubber Modifier.** Transportation Research Record. Washington, n.1339, p. 1-8.1992. USA, 1992.

HOLLERAN, G.; REED, J. R.[1] **Super Microsurfacings.** World Of Asphalt AAPA, Sydney. Taken from: www.slurry.com/techpapers/ techpapers_australia_ aapa.shtml Australia, 2000.

HOLLERAN, G.; REED, J. R.[2] **Emulsification of Asphalt Rubber Blends. In: Proceedings of the Asphalt Rubber** 2000 Conference p.309 to 318. Portugal, 2000.

HUANG, B.; MOHAMMAD, L. N.; GRAVES, P.S.; ABADIE, C. **Lousiana**

Experience with Crumb Modified Hot-mix Asphalt Pavement. Transportation Research Record. Washington, n.1789, p. 1-13. USA, 2002.

KIROSKI, D.; BURKE, B. AND PACKHAM, D. **Effect Of Substrate On Adhesion In Rubber Moulding**. Available at:<http:// staff.bath.cc.uk/mssdep/paper1.htm>. Accessed on: May 5, 1996.

MAMLOUK, M.S.; WOOD, L.E. AND GADALLAH, A. **A Laboratory Evaluation of asphalt emulsion mixtures by use of the Marshall and indirect tensile tests**. Transportation Research Board 754, p.17-22. USA, 1980.

NARUSCH, F.P. **Alaska Experience with Rubberized Asphalt Concrete Pavements**, 1979- 1982. Division of Design and Construction, Alaska Department of Transportation and Public Facilities, Central Region. USA, 1982.

PETROBRAS DISTRIBUIDORA S.A. **Cold premix (PMF)**. Available at: <http://www.br.com.br/portalbr/calandra.nsf#http://www.br.com.br/portalbr/calandra.nsf/0/09702B401A66490F03256DAD004A7381?openDocument&sasfalto> Accessed on: June 10, 2006.

PETROBRAS DISTRIBUIDORA S.A; ABEDA, Associação Brasileira das Empresas Distribuidoras de Asfaltos. **Asphalt emulsions, diluted asphalt and foam asphalt**, Block 2. Available at: http://www.cpgec.ufrgs.br/petcivil/Download/Apoio%20adidatico/Blo co%202-aula%208- %20Emulsao%20asfaltica%20asfalto%20diluido%20asfalto.pdf. Accessed on: March 6, 2008.

PiNHEiRO, J.H.M. E SOARES, J.B. **The Effect of Crumb Rubber Gradation and inderrubber Interaction Time on the Mechanical Properties of Asphalt-rubber Mixtures (Dry Process)**. in: Proceedings of the Asphalt Rubber 2003 Conference p.707 a 718. Brazil, 2003.

NOHARA, J.J.; ACEVEDO, C. R.; PiRES, B.C.C.; CORSiNO, R.M. GS-40 -

Solid Waste: Environmental Liabilities and Tire Recycling. Thesis, Sâo Paulo, 2005.

ODA, S.; NASCIMENTO, L.A.H; EDEL, G. **Application of Asphalt-Rubber in Bahia**. Brazilian Institute of Oil and Gas - IBP, 3rd Brazilian Congress of R&D in Oil and Gas. Bahia, 2005

ODA, Sandra, **Reuse of tires as an alternative for increasing the useful life of landfills**, Universidade Estadual de Maringà, 2002.

SANTANA, H. **Manual de Pré-Misturados a Frio**. IBP, Brazilian Petroleum Institute. Brazil, 1993.

SILVA JUNIOR, C. A.P.; SÓRIA, M.H.A. **Influences of moisture loss before compaction in cold sand-asphalt mixtures made with slow-setting cationic asphalt emulsion**. National Panorama of Transportation Research 2000. In: XIV Congresso de Pesquisa e Ensino em Transportes. p. 485 to 496 Brazil, 2000.

SINGLETON, T.M.; AIREY, G.D.; WIDYATMOKO, I. AND COLLOP, A.C. (2000) **Residual Bitumen characteristics following Dry Process Rubber-bitumen Interaction**. Proceedings of the Asphalt Rubber 2000 Conference p.309 to 318. Portugal, 2000.

SOARES, V. **ANIP reaches the milestone of 100 million old tires destroyed**. Available at: <http://www.anip.com.br/>. Accessed on: August 25, 2007.

SPECHT, L. P. **Evaluation of Asphalt Mixtures with the Incorporation of Recycled Tire Rubber.** Doctoral thesis. Federal University of Rio Grande do Sul. Porto Alegre, 2004.

TAKALLOU, H. B.; HICKS, R. G. **Development of improvement mix and construction guidelines for rubber-modified asphalt pavements**. Transportation Research Record. Washington, n. 1171, p. 113-120, USA, 1988.

THANAYA, I N.A. **Utilization of Waste Materials for Cold Asphalt Mixes**. Taken from the website: www.bibcouncil.de/ISSM2002/Proceedings/paperPDF/e02.pdf U.K., 2002.

TRELOAR, L.R.G. **The Physics of Rubber Elasticity**. 3rd Edition. Oxford. USA, 1975.

WALLER JR, H.F. **Emulsion Mix Design Methods: An Overview Transportation** Research Board 754, p.1-9. USA, 1980.

WAY, G. B. OGFC Meets **CRM Where the Rubber meets the Rubber 12 Years of** Durable Success. In: Proceedings Of The Asphalt Rubber 2000 Conference p. 15-31. Portugal, 2000.

Printed by Books on Demand GmbH, Norderstedt / Germany